SpringerBriefs in Electrical and Computer Engineering

Series Editors

Woon-Seng Gan, School of Electrical and Electronic Engineering, Nanyang Technological University, Singapore, Singapore

C.-C. Jay Kuo, University of Southern California, Los Angeles, CA, USA

Thomas Fang Zheng, Research Institute of Information Technology, Tsinghua University, Beijing, China

Mauro Barni, Department of Information Engineering and Mathematics, University of Siena, Siena, Italy

SpringerBriefs present concise summaries of cutting-edge research and practical applications across a wide spectrum of fields. Featuring compact volumes of 50 to 125 pages, the series covers a range of content from professional to academic. Typical topics might include: timely report of state-of-the art analytical techniques, a bridge between new research results, as published in journal articles, and a contextual literature review, a snapshot of a hot or emerging topic, an in-depth case study or clinical example and a presentation of core concepts that students must understand in order to make independent contributions.

More information about this series at http://www.springer.com/series/10059

Osama Bazan · Baha Uddin Kazi ·
Muhammad Jaseemuddin

Beamforming Antennas in Wireless Networks

Multihop and Millimeter Wave Communication Networks

 Springer

Osama Bazan
Ryerson University
Toronto, ON, Canada

Baha Uddin Kazi
Ryerson University
Toronto, ON, Canada

Muhammad Jaseemuddin
Ryerson University
Toronto, ON, Canada

ISSN 2191-8112   ISSN 2191-8120   (electronic)
SpringerBriefs in Electrical and Computer Engineering
ISBN 978-3-030-77458-5   ISBN 978-3-030-77459-2   (eBook)
https://doi.org/10.1007/978-3-030-77459-2

This Springer imprint is published by the registered company Springer Nature Switzerland AG
The registered company address is: Gewerbestrasse 11, 6330 Cham, Switzerland

Contents

Chapter 1
Introduction

Abstract Wireless networks are facing an ever-growing demand for high capacity, better coverage and support of new applications and broad range of services. Non-traditional wireless communications has recently surged to fulfill these increasing requirements. Multi-hop wireless networks extend the wireless coverage, improve the overall capacity and enable network auto-configuration with no infrastructure support. Millimeter wave (mmWave) communications exploit the less-congested spectrum, where available bandwidth is enormous. Both multi-hop and mmWave communications face some challenges that limit their benefits. To that end, "beam-forming antennas" is a promising technology to be utilized in conjunction with multi-hop and mmWave communications to unlock their full potential. However, it is insufficient to just plug-and-play a beamforming antenna system as it needs to be appropriately controlled by upper layers of the networking protocols. In this chapter, we detail the potential benefits of beamforming antennas and discuss the new challenges introduced by directional communication on the networking protocols originally designed with omni-directional antenna in mind.

1.1 Wireless Networks

The proliferation of computing devices in different form factors in conjunction with the rapid advances in communication and wireless networking has brought about a revolution in information technology. The wireless technologies provide a highly flexible, mobile, cost-effective and easy to use communication service that has started the era of pervasive communications and computing. Motivated by numerous applications and broad range of services, the research community is developing methods, architectures and protocols to overcome the challenges of wireless networks.

Traditionally, wireless networks are designed to provide single hop connectivity either to cellular base stations or Wireless Local Area Network (WLAN) access points. However, the possibility to extend the wireless coverage, improve the overall capacity and enable network auto-configuration with no infrastructure support has sparked the idea of multi-hop wireless networks in which nodes are able to forward data targeted for other nodes [4, 16]. The concept of multi-hop wireless networks

1
O. Bazan et al., *Beamforming Antennas in Wireless Networks*,
SpringerBriefs in Electrical and Computer Engineering,
https://doi.org/10.1007/978-3-030-77459-2_1

dates back to the 1970s with the introduction of packet radio networks. However, the development of the multi-hop wireless networking paradigm has recently surged with the increasing interest in Mobile Ad-hoc Networks (MANETs) and their applications in battlefield and disaster relief environments which evolved to a broader arena that encompasses wireless mesh networks, wireless sensor networks, wireless personal area networks, mobile multi-hop relay networks, multi-hop cellular networks, delay tolerant networks, and vehicular ad-hoc networks. The research on multi-hop wireless networks has attracted both academia and the wireless industry resulting in rapid commercialization, such as community mesh networks and WiMAX, as well as recent standardization efforts such as IEEE 802.11s and IEEE 802.16j.

Wireless networks are commonly deployed in the microwave spectrum, mainly below 6 GHz. As traffic demands are increasing at a rate that has never seen before, the scarcity of available bandwidth, in this range, has become a bottleneck for meeting the forthcoming capacity requirements. To overcome this hurdle, there has been a growing interest in exploiting the less-congested non-traditional spectrum, where available bandwidth is enormous, known as millimeter wave (mmWave) spectrum [15, 19, 21, 26]. The mmWave portion of the electromagnetic spectrum lies between 30 and 300 GHz (corresponding to wavelength from 10 to 1 mm), however, the industry has loosely considered mmWave to include frequencies above 10 GHz [19]. It has been estimated that mmWave provides abundant bandwidth that is more than 200 times the spectrum currently allocated below 3 GHz [15]. To exploit the benefits of mmWave communications, many standardization activities have been made in the recent years such as IEEE 802.15.3c [9] for WPAN, IEEE 802.11ad [10] for WLAN and 3GPP Release 15/16 [24] for 5G New Radio (NR). To complement these efforts the FCC has opened up more spectrum for use in 28 GHz/39 GHz and 60 GHz for licensed and unlicensed use respectively [5].

1.2 Beamforming Antennas: A Promising Technology

Motivated by the rapid deployment and emerging applications, the research community is interested in developing innovative solutions to address the challenges facing multi-hop wireless networks. Some of the key challenges include interference-limited capacity, power efficiency, quality of service and security. In this context, the "smart beamforming antennas" offer a promising technology to be utilized with multi-hop wireless networks [17, 27]. Although smart beamforming antennas have provided significant improvements in expanding coverage and mitigating interference when deployed in cellular networks [14], omni-directional antennas are still dominating all forms of multi-hop wireless networks. This is mainly due to the cost and size limitations. However, the recent advances in the antenna technology along with the shift towards higher operating frequencies have made it feasible to use this technology even in small, mobile and battery-operated devices [6, 12]. By the smart beamforming antenna technology, we refer to the antenna arrays technology coupled with sophisticated signal processing techniques responsible for the smart beamforming of the antenna radiation pattern.

Despite the potential of mmWave communication in overcoming bandwidth constraints, it poses new challenges as their propagation characteristics differ from those of the sub-6 GHz [22]. In particular, mmWave communication suffers from severe attenuation due to absorption by atmospheric oxygen, higher shadowing effect due to poor diffraction at higher frequencies as well as much higher penetration loss [22, 26]. To overcome these losses, large-scale antenna systems (also known as Massive MIMO) and beamforming antennas are considered the key enablers for mmWave communications [1, 7]. Recent studies have shown that directivity gain of smart beamforming antennas make mmWave communication feasible [20, 25].

1.2.1 Potential Benefits

Recently, the use of smart beamforming antennas in multi-hop wireless networks has received increasing attention in the research community due to their potential benefits and numerous advantages compared to omni-directional antennas. Some of these benefits include the following aspects:

- Since a directional antenna is able to radiate energy in the direction of the intended receiver, this transmission does not interfere with neighboring nodes residing in other directions. This increases the spatial reuse of the wireless channel as multiple simultaneous transmissions can take place within the same vicinity. The possibility of having more simultaneous transmissions in the same region promises a significant improvement in the wireless network capacity. Figure 1.1 shows this intuitive benefit and how the limited scope of the directional transmissions can increase the channel utilization significantly.
- The directional reception and the ability of sophisticated smart beamforming antennas to completely suppress the reception from interfering directions can significantly reduce interference, which deals with a major problem in multi-hop wireless networks.
- Focusing more energy in the intended direction (directional gain) increases the Signal-to-Noise Ratio (SNR). This improves the link reliability and robustness.
- For the same transmit power as omni-directional antennas, the directional gain of smart beamforming antennas is translated to communication range extension. This may lead to fewer-hops routes and consequently a reduction in the end-to-end delay. In addition, the communication range extension makes it possible to bridge network partitions improving the network connectivity.
- Reductions in the power consumption can trade-off the benefit of range extension. For a specific pair of nodes, smart beamforming antennas are able to reduce the transmit power while maintaining the same wireless link quality as omni-directional antennas. This makes smart beamforming antennas an attractive option to be used in battery-operated networks.

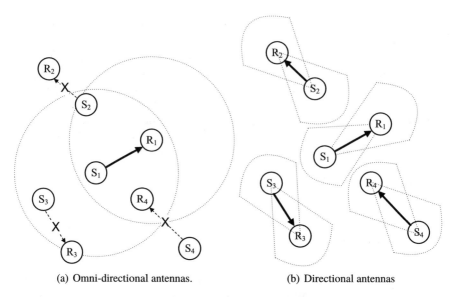

(a) Omni-directional antennas.   (b) Directional antennas

Fig. 1.1 An illustrative example of the spatial reuse benefit of directional antennas. In case of omni-directional antenna, a single communication S_1-R_1 precludes all other communications that involve the neighbors of either S_1 or R_1, whereas using directional antenna all four pair wise communications can occur simultaneously

- The unique features of smart beamforming antennas reduce the risks of eavesdropping and jamming, hence, providing more secure wireless communication.

In addition to the above-mentioned benefits, smart beamforming antennas provide more opportunities such as location estimation and efficient broadcasting.

1.2.2 Challenges in the Upper Layers

It is apparent that smart beamforming antennas offer a number of potential benefits for improving the performance of multi-hop wireless networks. Based on the network's usage and applications, the smart beamforming antenna can be efficiently utilized to satisfy the network's requirements such as higher throughput, better connectivity, additional quality of service guarantees and power savings just to name a few. However, it is not sufficient to plug-and-play a smart beamforming antenna to exploit the offered potential [13]. The smart beamforming antenna system needs to be appropriately controlled by upper layers of the networking protocol stack [18]. Antenna-aware schemes such as Medium Access Control (MAC) and routing need to be designed in order to realize the full potential of the smart beamforming antennas.

An important measure of a wireless network's operational performance is its capacity. The increasing use of multi-hop wireless networks and the ever-growing demand of bandwidth-intensive network applications are the driving force behind exploring innovative techniques that enable multi-hop wireless networks to provide higher throughput. Researchers have looked into solutions that can exploit the physical layer capabilities such as multi-channel networks, power control and rate control. In this book, we focus on the use of smart beamforming antennas in multi-hop wireless networks in overcoming their capacity limitations. By employing directional transmission and reception, the network capacity can be substantially enhanced. Several theoretical [8, 23, 28] and experimental studies [11, 18] have proven the intuitive benefits.

However, the conventional network protocols fail to interact with an underlying smart beamforming antenna since these protocols were originally designed to run on nodes equipped with omni-directional antennas. Lack of appropriate control over antenna beamforming may deteriorate the overall performance even below the level achieved by omni-directional antennas. This is the motivation behind our focus on the networking aspects of multi-hop wireless networks using smart beamforming antennas. Although, there is an adequate work done that deals with the issues and techniques of using smart beamforming antennas at the physical layer particularly in the context of cellular networks, the research in the area of multi-hop wireless networks is still ongoing. It is important to design innovative mechanisms, at the MAC and network layers, that are capable of harnessing the full potential of using smart beamforming antennas which is covered in this book.

The benefits of utilizing smart beamforming antennas are not free from trade-offs. Indeed, they pose new challenges. Although the directionality of the smart beamforming antennas promises a significant increase in the spatial reuse, it gives rise to a critical problem referred to as deafness [2]. Deafness was first identified in the context of a directional version of IEEE 802.11 MAC [3]. It occurs when a transmitter tries to communicate with a receiver but fails because the receiver is engaged in another communication in a different direction. Due to the characteristics of directional beamforming, the intended receiver is not able to receive the transmitter's signal and as a result it appears deaf to the transmitter. Networking protocols that are unaware with this new category of failure could result in a significant underutilization of the wireless channel leading to a significant loss in the network capacity. In addition to deafness, directional communication introduces other challenges such as new kinds of hidden terminal problems, head-of-line blocking and directional neighbor discovery. All the aforementioned challenges make the problem of exploiting the benefits of beamforming antennas in multi-hop wireless networks, at the MAC and network layers, an interesting research challenge.

To cope with the pressing need of running content-rich multimedia applications and real-time services, Quality of Service (QoS) support has become a vital component in today's wireless networks. Most of the exiting work on MAC and routing protocols for multi-hop wireless networks with beamforming antennas do not support QoS. However, the deployment of beamforming antennas could significantly spare the network resources that can be utilized for additional QoS guarantees. This motivates us to address the issue of QoS assured routing in this book.

1.3   Organization of the Book

In Chap. 2, we present a brief overview of beamforming antennas. We discuss some basic concepts of the antenna theory with focus on smart antenna technologies. In particular, we compare different types of beamforming antennas in terms of their capabilities and implementation complexity.

In Chap. 3, we give a brief review of conventional MAC and routing protocols for multi-hop wireless networks with omni-directional antennas. We then investigate the challenges and problems that face these protocols when beamforming antennas are employed.

In Chap. 4, we investigate the main design choices for directional MAC protocols. We classify the existing directional MAC protocols and overview some representative protocols in each class. We discuss the operation of three major directional MAC protocols and compare their performance using computer simulations.

In Chap. 5, we provide a brief overview of mmWave communication focusing on WLAN. We discuss some novel techniques adopted in IEEE 802.11ad and IEEE 802.11ay standards. We also present several MAC protocols that are designed to overcome the challenges of the of directional mmWave communications.

In Chap. 6, we discuss the mechanisms associated with directional routing. We also present several directional routing schemes that are designed for multi-hop wireless networks with beamforming antennas.

In Chap. 7, we develop a framework to analyze conflicts between wireless links in the presence of contention-based directional MAC protocols. Using our framework, we formulate the directional QoS routing problem as an optimization problem and present a joint routing and admission control algorithm to find single-path bandwidth-guaranteed routes.

In Chap. 8, several open research topics are presented and discussed.

References

1. Busari SA, Huq KMS, Mumtaz S, Dai L, Rodriguez J (2018) Millimeter-wave massive MIMO communication for future wireless systems: a survey. IEEE Commun Surv Tutor 20(2):836–869
2. Choudhury R, Vaidya N (2004) Deafness: a MAC problem in ad hoc networks when using directional antennas. In: IEEE international conference on network protocols (ICNP), Berlin, Germany, pp 283–292
3. Choudhury R, Yang X, Ramanathan R, Vaidya N (2002) Using directional antennas for medium access control in ad hoc networks. In: ACM international conference on mobile computing and networking (Mobicom), Atalanta, Georgia, pp 59–70
4. Conti M, Giordano S (2007) Multihop ad hoc networking: the reality. IEEE Commun Mag 45(4):88–95
5. Federal Communications Commission (2016) Use of spectrum bands above 24 GHz for mobile radio services. GN Docket No. 14–177, FCC 16–89
6. Giorgetti G, Cidronali A, Gupta SK, Manes G (2007) Exploiting low-cost directional antennas in 2.4 GHz IEEE 802.15.4 wireless sensor networks. In: European conference on wireless technologies, Munich, Germany, pp 217–220

7. Han S, Chih-Lin I, Xu Z, Rowell C (2015) Large-scale antenna systems with hybrid analog and digital beamforming for millimeter wave 5G. IEEE Commun Mag 53(1):186–194
8. Huang X, Wang J, Fang Y (2007) Achieving maximum flow in interference-aware wireless sensor networks with smart antennas. Elsevier Ad Hoc Netw 5(6):885–896
9. IEEE Standards Association: 802.15.3c-2009 - IEEE Standard for Information technology–Local and metropolitan area networks–Specific requirements–Part 15.3: Amendment 2: Millimeter-wave-based Alternative Physical Layer Extension. IEEE (2009)
10. IEEE Standards Association: IEEE Std 802.11 ad-2012, Part 11: Wireless LAN medium access control (MAC) and physical layer (PHY) specifications, amendment 3: enhancements for very high throughput in the 60 GHz band. IEEE Computer Society (2012)
11. Kohmura N, Mitsuhashi H, Watanabe M, Bandai M, Obana S, Watanabe T (2008) Unagi: a protocol testbed with practical smart antennas for ad hoc networks. ACM SIGMOBILE Mobile Comput. Commun Rev 12:59–61
12. Leang D, Kalis A (2004) Smart SensorDVBs: sensor network development boards with smart antennas. In: IEEE international conference on communications, circuits and systems (ICC-CAS), Chengdu, China, pp 1476–1480
13. Li G, Yang LL, Conner WS, Sadeghi B (2005) Opportunities and challenges for mesh networks using directional antennas. In: IEEE workshop on wireless mesh networks (WiMesh), Santa Clara, California
14. Liberti JC, Rappaport TS (1999) Smart antennas for wireless communications. Prentice Hall, NJ
15. Niu Y, Li Y, Depeng J, Li S, Vasilakos AV (2015) A survey of millimeter wave communications (mmWave) for 5G: opportunities and challenges. Wireless Netw 21:2657–2676
16. Perkins CE (2001) Ad Hoc networking. Addison-Wesley
17. Ramanathan R (2001) On the performance of ad hoc networks with beamforming antennas. In: ACM international symposium on mobile ad hoc networking and computing (MobiHoc), Long Beach, California, pp 95–105
18. Ramanathan R, Redi J, Santivanez C, Wiggins D, Polit S (2005) Ad hoc networking with directional antennas: a complete system solution. IEEE J Sel Areas Commun 23(3):496–506
19. Rangan S, Rappaport TS, Erkip E (2014) Millimeter-wave cellular wireless networks: potentials and challenges. Proc IEEE 102(3):366–385
20. Rappaport TS, Sun S, Mayzus R, Zhao H, Azar Y, Wang K, Wong GN, Schulz JK, Samimi M, Gutierrez F (2013) Millimeter wave mobile communications for 5G cellular: it will work!. IEEE Access 1:335–349
21. Rappaport TS, Heath RW Jr, Daniels RC, Murdock JN (2014) Millimeter wave wireless communications. Prentice Hall, NJ
22. Shafi M, Zhang J, Tataria H, Molisch AF, Sun S, Rappaport TS, Tufvesson F, Wu S, Kitao K (2018) Microwave vs. millimeter-wave propagation channels: key differences and impact on 5G cellular systems. IEEE Commun Mag 56(12):14–20
23. Spyropoulos A, Raghavendra CS (2003) Capacity bounds for ad-hoc networks using directional antennas. In: IEEE international conference on communications (ICC), Anchorage, Alaska, pp 348–352
24. Third Generation Partnership Project: 5G; NR; Base Station (BS) radio transmission and reception. 3GPP TS 38.104 Version 16.5.0 (2020)
25. Uwaechia AN, Mahyuddin NM (2020) A comprehensive survey on millimeter wave communications for fifth-generation wireless networks: feasibility and challenges. IEEE Access 8:62367–62414
26. Wang X, Kong L, Kong F, Qiu F, Xia M, Arnon S, Chen G (2018) Millimeter wave communication: a comprehensive survey. IEEE Commun Surv Tutor 20(3):1616–1653
27. Winters JH (2006) Smart antenna techniques and their application to wireless ad hoc networks. IEEE Wirel Commun Mag 13(4):77–83
28. Yi S, Pei Y, Kalyanaraman S (2003) On the capacity improvement of ad hoc wireless networks using directional antennas. In: ACM international symposium on mobile ad hoc networking and computing (MobiHoc), Annapolis, Maryland, pp 108–116

Chapter 2
Overview of Beamforming Antennas

Abstract In this chapter, we provide a concise overview on beamforming antennas. This chapter is devoted for interested readers from the networking community who do not have background related to this technology. The goal is to provide sufficient information that can help the reader fully understand the MAC and routing research reported in the subsequent chapters. It is worthy to note that we do not intend here to discuss all the physical layer aspects of the beamforming antenna technology. Rather, we present some fundamentals of the antenna theory, categorization, comparison of different variants of smart antennas and in particular the beamforming antennas technology of both sub 6 GHz and mmWave. Readers who are interested in additional details are referred to [2, 3, 8].

2.1 Fundamentals

The primary function of any radio antenna is to couple electromagnetic energy from one medium to another. There are few basic categories of antennas. The isotopic antennas are ideal radiating elements that can radiate power equally in all directions. The isotropic antenna is only used as a reference since it cannot exist in reality. Practically, omni-directional antennas are able to radiate energy equally in all directions in the azimuthal plane. The omni-directional antennas are simple and relatively cheap. Unlike omni-directional antennas, directional antennas are able to radiate the electromagnetic energy in one direction more than the others. Figure 2.1 presents a classification of antennas.

A fundamental principle in antenna theory is the reciprocity theorem. Based on this theorem, the transmission and reception characteristics of the antenna are the same. For instance, a directional antenna, that radiates more energy to a particular direction, receives more energy from the same direction that the others.

An important characteristic of an antenna is its gain as it is used to quantify the directionality of the antenna. The gain of an antenna in a certain direction indicates the relative power in that direction compared to the isotropic antenna. The gain is usually measured in dBi with the gain of an omni-directional antenna = 0 dBi.

© The Author(s), under exclusive license to Springer Nature Switzerland AG 2021   9
O. Bazan et al., *Beamforming Antennas in Wireless Networks*,
SpringerBriefs in Electrical and Computer Engineering,
https://doi.org/10.1007/978-3-030-77459-2_2

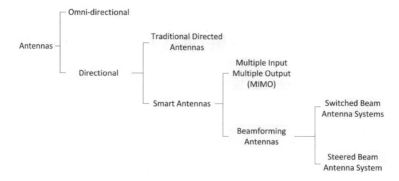

Fig. 2.1 Classification of antennas

The radiation pattern of the antenna is a representation of the gain values in all directions of space. For easier interpretation, the radiation pattern is usually plotted as a 2-D plot. Figure 2.2 shows an example for a radiation pattern of a directional antenna. A directional antenna pattern usually consists of a high gain main lobe (beam) and smaller gain side and back lobes. The peak gain is the maximum gain over all directions and lies along the axis of the main lobe which is also known as the boresight of the antenna. On the other hand, the null is a direction with very low gain. The radiation pattern usually have more than one null.

Another characteristic of a directional antenna is its beamwidth which formally refers to the angle subtended by the directions on either side of the boresight which are 3 dB less in gain. It is worthy to note that the gain and the beamwidth are related. Typically, the more directional the antenna, the higher its gain and the narrower its beamwidth.

Due to the complexity of realistic antenna radiation patterns, model abstraction is usually considered in upper-layers analysis and/or simulations. For example, an ideal directional antenna radiation pattern has a constant peak gain inside the main lobe and nulls outside it. To take side lobes into consideration, they can be abstracted by a low-gain constant circular arc outside the main lobe [9].

The relation between the antenna characteristics and the transmitted and received power is governed by Friss equation [2]. The received power P_r at a distance r from a transmitter with transmission power P_t is given by:

$$P_r = \frac{P_t G_t G_r}{K r^\delta},\tag{2.1}$$

where G_t and G_r are the transmitter and receiver gains along the straight line joining the transmitter and receiver, δ is the path loss exponent and K is a constant that is a function of the wavelength. A receiver can interpret the received signal if the received power is greater than or equal to the receiver sensitivity threshold.

Fig. 2.2 Antenna radiation
pattern with a main lobe
pointing at 0° and side lobes
with smaller gains

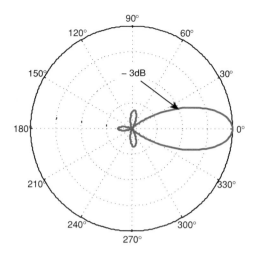

Based on (2.1), we can formally identify some of the benefits of directional antennas. Since the gain along the main beam of the directional antenna ($G_d > 1$) is always
greater than the gain of the omni-directional antenna ($G_o = 1$), the directional transmission results in communication range extension, denoted by r_{max}, for constant
transmit power P_t. Moreover, the gain from directional reception in conjunction
with the directional transmission can increase the communication range further. For
a constant distance r between the transmitter and the receiver, the directional transmission and/or reception will increase the received power P_r, hence increasing the
reliability (i.e. SNR) of the wireless link. Also, we can significantly decrease the
transmit power P_t while preserving the same received power P_r at a particular distance r. On the other hand, since the gain outside the main beam of the directional
antenna is very small (ideally null), the directional transmission reduces the generation of interference towards unintended directions. Similarly, the reception from one
direction suppresses the interference coming from other directions. This enhances
the spatial reuse significantly.

Directional antennas are often realized by means of antenna arrays. The antenna
array consists of identical radiating elements (e.g. dipoles) that are arranged in a particular geometric shape with different weights applied to the currents running through
them. The physical separation between the array elements is in terms of a fraction of
the wavelength. Hence, the size of the antenna array is inversely proportional to the
operating frequency. The overall radiation pattern of the antenna array is determined
by the number of elements, the element spacing, the geometrical configuration of
the array and the amplitude and phase of the applied signal to each element.

2.2 Smart Antennas Technology

The "smart antennas" technology refers to the combination of an antenna array with Digital Signal Processing (DSP) techniques and sophisticated antenna array control algorithms. These signal processing techniques allow the antenna elements to transmit and receive in an adaptive, spatially sensitive manner. In general, the term "smart antennas" represents a broad variety of adaptive antennas. However, there are two major variants of this technology which are the beamforming antennas[1] and the Multiple Input Multiple Output (MIMO) systems [6].

2.2.1 Beamforming Antennas

The beamforming antenna system employs DSP techniques to adaptively and intelligently change the radiation pattern of the antenna array. In particular, the DSP algorithms are used to estimate the Direction-of-Arrival (DoA) of the signal and use this information to calculate the weights applied to the signal at each antenna element. The main goal is to maximize the SNR towards the corresponding transceiver. The amount of control over this beamforming process relies on the sophistication of the applied algorithms.

2.2.2 MIMO Systems

MIMO systems represent the most sophisticated technology under the umbrella of smart antennas. A MIMO link utilizes adaptive antenna arrays at both the transmitter and the receiver to overcome the limitations of multi-path environments. There are two operational modes: spatial multiplexing and diversity. Spatial multiplexing gain is achieved when the multiple independent streams of data are transmitted out of different antennas with equal power. Each transmitted stream has a different spatial signature due to the rich multipath environment. These differences are exploited by the receiver signal processor to separate the streams. On the other hand, when dependent streams are transmitted, the rich multipath can help the data streams fade independently at the receiver and diversity gain is achieved.

---

[1]In the context of multi-hop wireless networks, many refer to "beamforming antennas" as "directional antennas". In this book, we will use the two terms interchangeably.

2.3 Types of Beamforming Antennas

Smart beamforming antennas are classified into switched beam systems and steered beam systems.

2.3.1 Switched Beam Antenna Systems

In switched beam systems, the antenna array is combined with a fixed Beam Forming Network (BFN). The BFN consists of a predetermined set of weight vectors, where the configuration of weights in a vector determines the direction in which the antenna radiation pattern is beamformed. Based on the direction-of-arrival estimation, the BFN chooses a weight vector to be applied to the signal received/transmitted by the antenna array. In other words, the antenna adaptively switches to one of the predefined set of beams as shown in Fig. 2.3.

Switched beam antennas can provide most of the benefits of smart antennas at a small fraction of complexity and expense. Spatial reuse, range extension and power saving are possible with this type of smart antennas. However, they do not guarantee maximum gain due to scalloping. Scalloping is the roll-off of the antenna pattern as a function of the angle from the boresight. If the desired direction is not on one of the predetermined boresights, the transceiver will suffer from gain reduction. Moreover, switched beam antennas are not able to fully eliminate the interference outside the main lobe due to the absence of control on the side lobes.

Fig. 2.3 Switched beam antenna system has a predefined set of beams to choose from

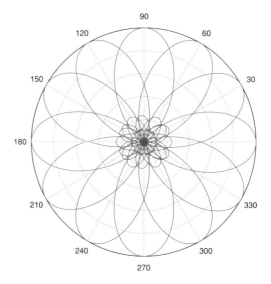

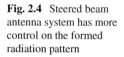
Fig. 2.4 Steered beam antenna system has more control on the formed radiation pattern

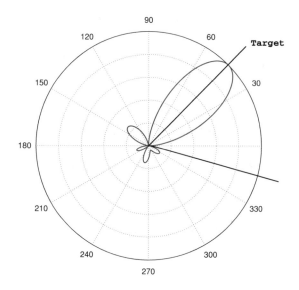

2.3.2 Steered Beam Antenna Systems

They are also known as adaptive antenna array systems. They provide a high degree of flexibility in configuring the radiation patterns. Using a variety of sophisticated signal processing algorithms, the adaptive array antennas can adapt their weights in order to maximize the resulting Signal to Interference and Noise Ratio (SINR). The boresight of the main lobe can be directed towards the target to optimize the gain. This type is known as phased antenna arrays. By increasing the complexity of the DSP algorithms, nulls can be additionally placed in the direction of interfering sources to suppress their interference. Figure 2.4 shows a possible radiation pattern formed by an adaptive antenna array system.

Although steered beam systems can outperform switched beam systems especially in multi-path environments, the associated complexity and cost are limiting factors [6]. The need to continuously locate and track various types of signals complicates the signal processing task which results in a significant increase in the power consumption.

2.4 Millimeter Wave Beamforming

In recent years wireless data traffic grows manifold due to the growing number of connected devices and wide range of bandwidth hungry applications. In order to meet this challenging demand, millimeter wave (mmWave) communication emerged to deploy high capacity data link, because of its wider spectral range and high data rate. Therefore, mmWave is considering an enabling technology for the successful

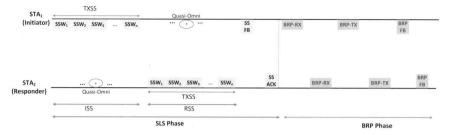

Fig. 2.5 Millimeter wave beamforming procedure

commercialization of next generation wireless networks (5G and beyond) to satisfy the requirements and user demand.

The gain of mmWave mostly depends on physical layer, medium access control layer and hardware implementation. Because of the small wavelength, to achieve the gain of mmWave, multiple antenna technology with beamforming communication is a key scheme. The use of directional multiple antenna beamforming at the transmitter and/or receiver can achieve multiplexing gain, diversity gain and antenna gain. In light of this, analog beamforming (ABF), digital beamforming (DBF), hybrid beamforming (HBF) and adaptive beamforming (AdBF) have been considering to facilitate the evolution of 5G and beyond communication. IEEE 802.11ad and IEEE 802.11ay support mmWave beamforming in the 60 GHz band [10, 11]. Because of its very short wavelength, 60 GHz mmWave band can implement highly directional antennas in a very small size. Therefore, to overcome the propagation loss and fading, beamforming antenna arrays in 60 GHz mmWave can improve the capacity and throughput significantly. In the classical systems, digital beamforming is a popular choice because of its greater flexibility to implement efficient beamforming algorithm. In this method a dedicated RF chain is required for each antenna element, which results high power consumption and complexity. As a result, DBF is not considered as a viable method for large scale mmWave antenna system. Low complexity ABF method have been adopted for short distance indoor mmWave communication in the 60 GHz band and HBF method is considering for outdoor mmWave communication [1, 7]. In this section we discuss the antenna beamforming mechanisms for mmWave communication.

Beamforming is a mechanism that is used by the transmitter and receiver to achieve the necessary directional multigigabit (DMG) link budget. It also enables the stations to train and refine the transmit and receive antennas for subsequent communication. The IEEE 802.11ad and IEEE 802.11ay specifies beamforming process in three phases: sector level sweep (SLS), beam refinement protocol (BRF) and beam tracking (BT) [4, 5, 7]. Figure 2.5 shows mmWave beamforming process [4, 7].

1. **Sector Level Sweep (SLS)**: The beamforming process starts with SLS from the initiator. This phase can include four subphases: initiator sector sweep (ISS), responder sector sweep (RSS), sector sweep feedback (SSFB) and sector sweep

ACK (SSACK). During the ISS, the initiator sends multiple transmit sector sweep (SSW) frames continuously changing the direction. Receiver antenna remains in the quasi-omnidirectional mode in this subphase. After the completion of ISS, responder sends SSW frames during the RSS period containing the best TX sector id received from the initiator in the ISS subphase. After the completion of RSS subphase, initiator sends the SSFB frame containing the best TX sector ID of both initiator and responder. Finally, SSFB frame acknowledged by the responder with SSACK frame. This is how the transmitter and receiver find the transmit sector with highest signal quality [4, 11].

2. **Beam Refinement Protocol (BRP)**: Once the best sector pair identified in the SLS phase, the BRP phase starts based on the request of initiator or responder. In this phase participating stations iteratively train its TX and RX antenna array(s) and improve its TX and RX antenna configuration. This phase comprises BRP setup subphase, an optimum multi sector ID (MID) detection subphase, beam combining (BC) subphase and beam refinement transaction subphase. The intend and capabilities to conduct some or all of these subphases exchange in the BRP setup subphase. The MID detection and BC subphases are optionally used to find better antenna weight vector (AWV). The beam refinement transaction subphase consist a set of beam refinement request and response frames appending transmit training (TRN-T) and receive training (TRN-R) subfield. This subphase can be used either by a transmit or a receive station or both stations to explore a border set of AWVs [4, 7].

3. **Beam Tracking (BT)**: After successful completion of the SLS and BRP phases, transmitter and receiver used the best refined beam for data transmission. The BT is an optional phase that employed during the data transmission to adjust channel changes. If the quality of a beamformed link goes down to a specified threshold, the beam tracking initiator can append receive training (TRN-R) field or transmit training (TRN-T) field to data frames to perform beam tracking. The BT phase find an optimal beam to continue seamless data transmission [1, 4].

The hardware implementation is not discussed in this section. In Chap. 5, we discuss about the enhanced medium access control (MAC) mechanism of millimeter wave directional multigigabit communication in details.

References

1. Abbas WB, Gomez-Cuba F, Zorzi M (2017) Millimeter wave receiver efficiency: a comprehensive comparison of beamforming schemes with low resolution ADCs. IEEE Trans Wirel Commun 16(12):8131–8146
2. Balanis CA (1997) Antenna theory: analysis and design. Wiley, NY
3. Balanis CA, Ioannides P (2007) Introduction to smart antennas. Morgan and Claypool Publishers
4. IEEE Standards Association (2012) IEEE Std 802.11 ad-2012, Part 11: wireless LAN medium access control (MAC) and physical layer (PHY) specifications, amendment 3: enhancements for very high throughput in the 60 GHz band. In: IEEE Computer Society

5. Jo O, Hong W, Choi ST, Chang SH, Kweon C, Oh J, Cheun K (2014) Holistic design considerations for environmentally adaptive 60 GHz beamforming technology. IEEE Commun Mag 52(11):30–38

6. Karthikeyan Sundaresan SL, Sivakumar R (2006) On the use of smart antennas in multi-hop wireless networks. IEEE international conference on broadband communications, networks and systems, California, San Jose, pp 1–10

7. Kutty S, Sen D (2015) Beamforming for millimeter wave communications: an inclusive survey. IEEE Commun Surv Tutor 18(2):949–973

8. Liberti JC, Rappaport TS (1999) Smart antennas for wireless communications. Prentice Hall, NJ

9. Ramanathan R (2001) On the performance of ad hoc networks with beamforming antennas. In: ACM international symposium on mobile ad hoc networking and computing (MobiHoc), Long Beach, California, pp 95–105

10. Rappaport TS, Shu S, Mayzus R, Zhao H, Wang K, George NW, Jocelyn KS, Mathew S, Felix G (2013) Millimeter wave mobile communications for 5G cellular: it will work. IEEE Access 1:335–349

11. Zhou P, Kaijun C, Xiao H, Xuming F, Yuguang F, Rong H, Yan L, Yanping L (2018) IEEE 802.11 ay-based mmWave WLANs: design challenges and solutions. IEEE Commun Surv Tutor 20(3):1654–1681

Chapter 3
MAC Issues

Abstract Conventional wireless MAC protocols, such as IEEE 802.11, implicitly assumes an omni-directional antenna at the physical layer. They were designed to overcome the well-known challenges of the wireless access such as hidden terminal problem and exposed terminal problem. With the deployment of beamforming antennas, simple modifications in IEEE 802.11 operation fail to exploit the potential benefits. The unique characteristics of beamforming antennas pose unprecedented challenges. In this chapter, we discuss the main beamforming-related challenges facing the medium access control in multi-hop wireless networks.

3.1 Medium Access Control with Omni-Directional Antennas

The wireless medium is open and shared by several nodes in the network. If acquiring this resource is left uncontrolled, multiple nodes may try to access it at the same time. The goal of the MAC protocol is to set the rules in order to enable efficient and fair sharing of the common wireless channel [3, 10]. The MAC protocol typically needs to maximize the channel utilization by having as many simultaneous communications as possible.

Medium access control protocols for wireless networks [12] may be classified into two major categories: contention-based and contention-free MAC. In contention-based MAC, nodes compete to access the shared medium through random access. In case of conflict occurrence, a distributed conflict resolution algorithm is use to resolve it. The most commonly considered contention-based MAC mechanism is the Carrier Sensing Multiple Access with Collision Avoidance (CSMA/CA). On the other hand, contention-free MAC is based on a controlled access in which the channel is allocated to each node according to a predetermined schedule.

The IEEE 802.11 Distributed Coordinated Function (DCF) is one of the CSMA/CA based protocols which has lately received a great attention due to its simplicity. In IEEE 802.11 DCF MAC [9], a node wishing to access the wireless medium should perform physical carrier sensing before initiating transmission. This is the CSMA part of the protocol. However, the performance of CSMA degrades significantly in

© The Author(s), under exclusive license to Springer Nature Switzerland AG 2021   19
O. Bazan et al., *Beamforming Antennas in Wireless Networks*,
SpringerBriefs in Electrical and Computer Engineering,
https://doi.org/10.1007/978-3-030-77459-2_3

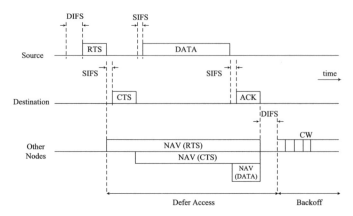

Fig. 3.1 Channel reservation in IEEE 802.11 MAC

multi-hop wireless networks due to the hidden terminal problem [10]. When two nodes are outside the carrier sensing range of each other, they are said to be hidden. If both nodes attempt to communicate with a common node, collision occurs at the receiving node. To overcome this problem, collision avoidance is implemented by a handshaking mechanism before data transmission [2]. The data transmission is preceded by transmitting a short Request-To-Send (RTS) packet to the intended receiver which in turn responds with a short Clear-To-Send (CTS) packet if the channel is idle at the receiver site for Short Interframe Spacing (SIFS) period. Both RTS and CTS packets contain the proposed duration of transmission. Nodes located in the vicinity of the communicating nodes, which overhear either of these control packets, must themselves defer transmission for the proposed duration. This is called Virtual Carrier Sensing (VCS) and is implemented through a mechanism called the Network Allocation Vector (NAV). A node updates the value of the NAV with the duration field specified in the RTS or CTS. Thus, the area covered by the transmission range of the sender and receiver is reserved. This procedure reduces the probability of collision dramatically. Figure 3.1 shows the collision avoidance operation in IEEE 802.11 MAC.

The IEEE 802.11 MAC protocol uses a backoff mechanism to resolve channel contention. Before initiating a transmission, each node performs both virtual and physical carrier sensing. If NAV is not set, and the channel is sensed idle, the node defers for DCF Interframe Spacing (DIFS) period before sending its packet. If the channel is found busy (by physical carrier sensing), the node chooses a random backoff interval from $[0, CW]$, where CW is called the contention window. The CW is initialized to the value of CW_{min}. After every idle slot time, the node decrements the backoff counter by one. When the counter reaches zero, the node can transmit its packet. In case a CTS or ACK packet is not received back, the node assumes a collision has occurred with some other transmission and it invokes the binary exponential backoff algorithm. In this backoff algorithm, the node doubles its CW, chooses a new backoff interval and tries retransmission again once the backoff timer expires. The CW is doubled on each collision until it reaches a maximum threshold, called CW_{max}. Retransmission retries are limited by a threshold after which the packet is discarded. If the medium is sensed busy during the backoff stage, the node freezes its backoff and resumes it once the medium has become idle for DIFS duration. Once a transmission is successfully transmitted, CW is initialized to its minimum value for the next transmission.

3.2 MAC Challenges with Beamforming Antennas

The design of IEEE 802.11 implicitly assumes an omni-directional antenna at the physical layer. When smart beamforming antennas are used, IEEE 802.11 MAC does not work properly. Researchers have looked into adapting IEEE 802.11 to the case of beamforming antennas. Choudhury et al. propose a directional version of IEEE 802.11 DCF MAC under the name of "Basic DMAC" [6] which is considered the benchmark for directional medium access control protocols.[1] To exploit the spatial reuse benefits, the Basic DMAC require the active nodes to perform carrier sensing, back-off, and the four-way handshake in a directional mode while the idle nodes reside in an omni-directional mode.

Conventional wireless MAC protocols were designed to overcome the challenges of the wireless medium such as hidden terminal problem and exposed terminal problem [12]. The unique characteristics of beamforming antennas pose unprecedented challenges that should be considered in the design of the directional MAC protocols for wireless ad hoc networks. In this section, we discuss the main beamforming-related challenges facing the medium access control.

3.2.1 Deafness

While exploiting the spatial reusability using beamforming antennas, deafness is by far the most critical challenge [8, 14]. Deafness [4] was first identified in the context of the Basic Directional MAC (DMAC) protocol [6]. It occurs when a transmitter tries to communicate with a receiver but fails because the receiver is beamformed towards a direction away from the transmitter. Due to the characteristics of directional beamforming, the intended receiver is unable to receive the transmitter's signal and as a result appears deaf to the transmitter.

Considering the example in Fig. 3.2, nodes B and C are engaged in directional communication while node A is in the backoff stage. Node A cannot sense the ongoing communication and is basically unaware of it, thus it attempts to communicate with node B at the end of its backoff. Since node B is beamforming in another

---

[1]The term "directional MAC protocols" is commonly used to refer to the MAC protocols designed particulary for wireless networks with beamforming antennas.

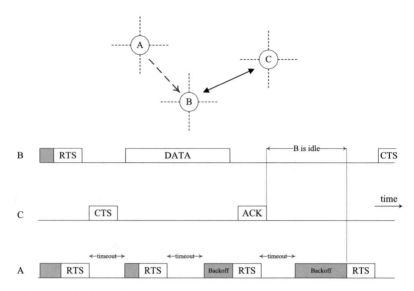

Fig. 3.2 A scenario illustrating the deafness problem

direction, it is deaf to node A's transmission and cannot respond. Due to the absence of CTS response, node A typically considers this kind of failure as an indication of collision and reacts accordingly. It invokes the binary exponential backoff algorithm before attempting retransmissions. Multiple retransmissions could happen until node B has finished the dialog with node C and switches back to the omni-directional mode. These unnecessary retransmissions reduce the network capacity. Moreover, the exponential increase in the backoff contention window results in channel underutilization as shown in Fig. 3.2.

The consequences of deafness may be even more severe. Assume that node C has multiple packets to send to node B. Once node C has finished transmitting the first packet, it immediately prepares to transmit the next packet by choosing a backoff interval from the minimum contention window. It is likely that node A is still engaged in the large backoff phase when node C finishes counting down its small backoff value for the second packet. Node C acquires channel access and communicates again with node B. This scenario can continue for a long time, causing node A to drop multiple packets before it gets fortunate enough to grab the channel access from node C. This scenario depicts that deafness may lead to short-term unfairness between flows that share a common receiver. If the MAC protocol requires the node to carrier-sense, backoff and communicate directionally, it may suffer from prolonged period of deafness if it has multiple back-to-back packets to be transmitted. Moreover, a chain of deafness is also possible in which each node attempting to communicate with a deaf node becomes itself deaf to another node. This could also result in a deadlock scenario [4].

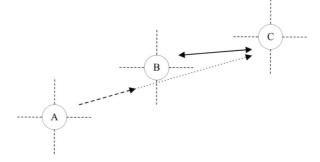

Fig. 3.3 A scenario to illustrate the hidden terminal problem due to the asymmetry in gain

3.2.2 New Hidden Terminals

The traditional hidden terminal problem in wireless networks occurs when two nodes are outside the carrier sensing range of each other and both of them attempt to communicate with a common node causing collision. The collision avoidance concept was proposed to solve this problem which is implemented by means of RTS/CTS handshaking before data transmission [2]. The RTS/CTS handshaking mechanism informs the neighboring nodes about imminent communication.

In the context of beamforming antennas, the hidden terminal problem occurs when a potential interferer could not receive the RTS/CTS exchange due to its antenna orientation during the handshake and then initiates a transmission that causes collision. There are two new types of directional hidden terminal problems [7].

3.2.2.1 Hidden Terminal Due to Asymmetry in Gain

This problem is basically due to the fact that the antenna gain in the omni-directional mode (G_o) is smaller than the gain when the antenna is beamformed (G_d). If an idle node is listening to the medium omni-directionally, it will be unaware of some ongoing transmissions that could be affected with its directional transmission.

To explain this type of hidden terminal problem, we refer to the scenario in Fig. 3.3. Assume that node A and node C are out of each other's range when one is transmitting directionally (with gain G_d) and the other is receiving omni-directionally (with gain G_o). However, they are within each other's range only when both the transmission and reception are done directionally (both with gain G_d). First, node B transmits RTS directionally to node C, and node C responds back with a directional CTS. Node A is idle (still in omni-directional mode) so it is unable to hear the CTS. Data transmission begins from node B to node C with both nodes pointing their transmission and reception beams towards each other. While this communication is in progress, node A has a packet to send to node B. Node A beamforms towards node B (which is the same direction of node C) and performs the carrier sensing. Since the channel is sensed idle, node A sends a directional RTS to node B. However,

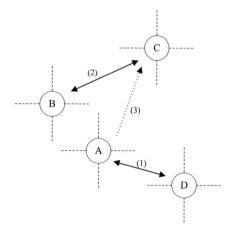

since node C is receiving data directionally using a beam pointed toward node B (and node A), the RTS from node A interferes with node B's data transmission at the receiver C causing collision.

3.2.2.2   Hidden Terminal Due to Unheard RTS/CTS

This type of hidden terminal problem occurs as a result of the loss in the channel state information during beamforming. When a node is involved in a directional communication, it would appear deaf to all other directions and important control packets may be lost during that time. In contrast to the deafness problem in which the packet cannot be received by its intended receiver, this type of new hidden terminals occurs when a "neighboring node" fails to receive the channel reservation packets (RTS/CTS) exchanged by a transmitter-receiver pair. Hence, it becomes unaware of the imminent communication between that particular transmitter-receiver pair and accordingly could later initiate a transmission that causes collision. An illustrating example is shown in Fig. 3.4. Suppose that node A is engaged in a directional communication with node D. While this communication is in progress, node B sends RTS to node C which in turns replies with CTS. Since node A is beamformed towards node D, it cannot hear CTS from node C. While the communication between node B and node C is in progress, node A finishes the communication with node D and now decides to transmit to node C. Since the DNAV at node A is not set in the direction of node C (due to the unheard CTS), node A transmits RTS to node C causing collision at node C.

Fig. 3.5 A scenario to
illustrate the head-of-line
blocking problem

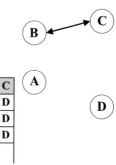

3.2.3 Head-of-Line Blocking

The Head-of-Line (HoL) blocking problem with directional MAC protocols was
first identified in [11]. It occurs as a result of the typically used First-In-First-Out
(FIFO) queueing policy. This policy works fine in the presence of omni-directional
antennas since all outstanding packets use the same medium. If the medium is busy, no
packets can be transmitted. However, in case of beamforming antennas, the medium
is spatially divided and it may be available in some directions but not others. If the
packet at the top of the queue is destined to a busy node/direction, it will block all
the subsequent packets even though some of them can be transmitted as illustrated
in Fig. 3.5. Using the FIFO queueing policy, although node A has packets that can
be transmitted to node D, they are blocked by the packet destined to the busy node
C. The HoL blocking problem is aggravated when the top packet goes into a round
of failed retransmissions including their associated backoff periods as discussed
in [1].

3.2.4 Communication Range Under-Utilization

In contrast to the previous problems that mainly offset the benefit of spatial reuse
introduced by beamforming antennas, the operation of the directional MAC protocol may limit the full exploitation of the communication range extension offered by
beamforming antennas. If the protocol requires the omni-directional transmission of
control packets or the idle node to reside in an omni-directional mode, the communication range is limited. It is possible for nodes to communicate over the extended
range if both the transmitter and the receiver could agree to beamform towards each
other at the same time which is a challenging issue in the presence of asynchronized
medium access. Following the terminology introduced in [6], the node has three
types of neighbors: (1) The Omni-Omni (OO) neighbors: Those are neighbors that
can only receive the omni-directional transmissions of the node when they are listening in an omni-directional mode. (2) The Directional-Omni (DO) neighbors: Those
are neighbors that can also receive the directional transmissions of the node when

they are listening in an omni-directional mode. (3) The Directional-Directional (DD) neighbors: Those are neighbors that can receive the directional transmissions of the node only if they are already beamformed in the direction of the node. The challenge facing the MAC protocols is how to allow communication to occur between DD-neighbors.

3.2.5 MAC-Layer Capture

Since a packet can be received from any direction, it is common that the antenna of an idle node resides in an omni-directional mode in order to be able to listen in all directions. When a signal is detected, the antenna will beamform towards the direction of maximum received power, receive the packet, decode it and pass it up to the MAC layer. If the packet is not destined to this node, the packet will simply be dropped. The time the node wastes in receiving packets, not intended to it, might refrain the node from transmitting/receiving useful packets to/from other directions thus resulting in channel underutilization. This problem is identified in [5] under the name of "MAC-layer capture" as a limiting factor in the potential increase in the spatial reuse when beamforming antennas are employed. It is worthy to note that the MAC-layer capture problem is not restricted to the use of beamforming antennas. However, in the case of omni-directional antennas, there is little motivation to avoid being captured by ongoing frames because captured nodes are not expected to initiate any concurrent transmissions until the medium is idle. On the contrary, beamforming antennas spatially divide the shared medium and a transmission in one direction does not affect other directions.

In the context of beamforming antennas, the MAC-layer capture problem basically occurs when the node does not perform any intelligent control of the underlying antennas when it is idle. Considering the example in Fig. 3.6, suppose that node A

Fig. 3.6 A scenario to illustrate the MAC-layer capture problem

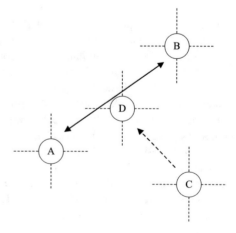

has a packet to transmit to node *B* and node *C* has a packet to transmit to node *D*. Using beamforming antennas, the two dialogues could occur concurrently. However, if node *A* starts its transmission first towards node *B*, the idle node *D* will get engaged in receiving node *A*'s transmission thus its concurrent communication with node *C* is not possible. The MAC-layer capture problem does not only reduce the spatial reuse but also leads to the negative consequences of deafness as pointed out in [13, 15].

References

1. Bazan O, Jaseemuddin M (2008) An opportunistic directional MAC protocol for multihop wireless networks with switched beam directional antennas. In: IEEE international conference on communications (ICC), Beijing, China, pp 2775–2779
2. Bharghavan V, Demers A, Shenker S, Zhang L (1994) MACAW: a media access protocol for wireless LANs. In: ACM international conference of the special interest group on data communication (SIGCOMM), London, UK, pp 212–225
3. Chandra A, Gummalla V, Limb JO (2000) Wireless medium access control protocols. IEEE Commun Surv Tutor 3(2):2–15
4. Choudhury R, Vaidya N (2004) Deafness: a MAC problem in ad hoc networks when using directional antennas. In: IEEE international conference on network protocols (ICNP), Berlin, Germany, pp 283–292
5. Choudhury R, Vaidya N (2007) MAC-layer capture: a problem in wireless mesh networks using beamforming antennas. IEEE sensor, mesh and ad hoc communications and networks (SECON), San Diego, California, pp 401–410
6. Choudhury R, Yang X, Ramanathan R, Vaidya N (2002) Using directional antennas for medium access control in ad hoc networks. In: ACM international conference on mobile computing and networking (Mobicom), Atalanta, Georgia, pp 59–70
7. Choudhury RR, Yang X, Ramanathan R, Vaidya NH (2006) On designing MAC protocols for wireless networks using directional antennas. IEEE Trans Mob Comput 5(5):477–491
8. Gossain H, Cordeiro C, Cavalcanti D, Agrawal DP (2004) The deafness problems and solutions in wireless ad hoc networks using directional antennas. In: IEEE global telecommunications conference (GLOBECOM) workshops, pp 108–114
9. IEEE (1999) IEEE 802.11 standard: wireless LAN medium access control (MAC) and physical layer (PHY) specification
10. Jurdak R, Lopes CV, Baldi P (2004) A survey, classification and comparative analysis of medium access control protocols for ad hoc networks. IEEE Commun Surv Tutor 6(1):2–16
11. Kolar V, Tilak S, Abu-Ghazaleh NB (2004) Avoiding head of line blocking in directional antenna. In: IEEE international conference on local computer networks (LCN), Zurich, Switzerland, pp 385–392
12. Kumar S, Raghavanb VS, Deng J (2006) Medium access control protocols for ad hoc wireless networks: a survey. Elsevier Ad Hoc Netw J 4(3):326–358
13. Li Y, Safwat AM (2006) DMAC-DACA: enabling efficient medium access for wireless ad hoc networks with directional antennas. In: IEEE international symposium on wireless pervasive computing (ISWPC), Phuket, Thailand, pp 1–5
14. Takata M, Bandai M, Watanabe T (2007) A MAC protocol with directional antennas for deafness avoidance in ad hoc networks. In: IEEE global telecommunications conference (GLOBECOM), Washington, USA, pp 620–625
15. Wang J, Zhai H, Li P, Fang Y, Wu D (2009) Directional medium access control for ad hoc networks. Wirel Netw (Springer) 15(8):1059–1073

Chapter 4
Directional MAC

Abstract It is not straightforward to leverage the potential benefits of smart beam-forming antennas in multihop wireless networks. Since traditional wireless MAC protocols were originally designed with the implicit assumption that all the nodes are equipped with omni-directional antennas, they are not capable of exploiting the advantages of beamforming antennas. In addition, the new challenges, associated with directional beamforming, can offset the gains if left unaddressed as discussed in the previous chapter. In this chapter, we first discuss the main design choices for directional MAC protocols. We then present a taxonomy of the directional MAC protocols and overview some representative protocols in each class. Finally, we explain in details the operation of three major directional MAC protocols and present some simulation results to evaluate their performances.

4.1 Directional MAC Design Choices

Due to the wide spread usage of multi-hop wireless networks with omni-directional antennas, researchers have focused on tweaking some steps of the well-established MAC protocols to accommodate the new environment. In this section, we discuss several design choices for directional MAC along with their advantages and disadvantages.

4.1.1 Transmission Mode of RTS/CTS Packets

A common design choice for a directional MAC is to exchange DATA and ACK packets directionally to exploit the benefits of beamforming antennas. However, there are many variations in the transmission mode of RTS/CTS packets when directional CSMA/CA based protocols are considered. Figure 4.1 shows the coverage range of the different transmission modes.

Like IEEE 802.11, RTS/CTS can be transmitted while the antenna is in the omni-directional mode [14, 32, 33]. The transmission of RTS/CTS in all directions is

O. Bazan et al., *Beamforming Antennas in Wireless Networks*,
SpringerBriefs in Electrical and Computer Engineering,
https://doi.org/10.1007/978-3-030-77459-2_4

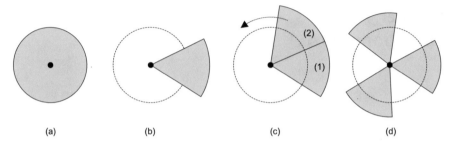

Fig. 4.1 The coverage range of different tranmission modes. **a** Omni-directional. **b** Uni-directional. **c** Multi-directional sequential. **d** Multi-directional concurrent

beneficial to inform all neighbors about imminent communication and hence, reduce the instances of deafness and hidden terminal problem significantly. Moreover, the communicating node does not need to know/cache the beamforming information of other nodes. However, this conservative reservation scheme is not commonly used since it comes on the expense of reduced spatial reuse and limited communication range which are the main benefits of using beamforming antennas.

To enable more simultaneous transmissions and extended range communications, several directional MAC protocols [3, 10, 24, 42] in the literature have employed directional RTS/CTS handshakes for channel reservation. However, the beamforming information of the destination node should be known a priori at the source node which is indeed a challenging task. Moreover, this aggressive reservation scheme lends itself to deafness and directional hidden terminal problems. Several performance studies [18, 49] have shown that while directional RTS/CTS may result in more transmission failures, this scheme outperforms the omni-directional RTS/CTS scheme in terms on enhanced throughput and reduced delay. However, a performance analysis that takes deafness into consideration [4] shows that these results are applicable only at certain range of beamwidths. For narrow beamwidths, the negative impact of deafness offsets the benefits of spatial reuse and results in a steep decrease in the saturation throughput.

To address the fundamental tradeoff between omni-directional RTS/CTS and directional RTS/CTS, some MAC protocols [16, 27, 43] rely on the multi-directional RTS/CTS for channel reservation. With switched beam antennas, the control packets are transmitted circularly, one direction after the other, to allow for collision and/or deafness avoidance. If one direction is sensed busy, it is usually skipped and the RTS/CTS transmission continues on the rest of the directions. Hence, the imminent communication is not blocked unnecessarily as in the omni-directional RTS/CTS case. The main drawback of this scheme is its large control overhead that can sometimes offset the spatial reuse benefits [11].

Few proposed MAC protocols [2, 5] consider the transmission of the RTS/CTS packets concurrently instead of sequentially to multiple directions. This scheme requires the use of sophisticated adaptive antenna array systems that can form multi-beam radiation patterns.

4.1.2 *Virtual Carrier Sensing*

As mentioned before, virtual carrier sensing is used in CSMA/CA to protect an ongoing transmission from being lost due to hidden terminals. It is implemented by maintaining a Network Allocation Vector (NAV) at each node. NAV contains a defer timer which indicates how long the channel will remain busy and is updated based on the duration field of overheard packets. In an omni-directional MAC, the virtual carrier sensing applies to all directions. However, with beamforming antennas, the channel is spatially divided and each node has to maintain a directional NAV (DNAV) table as proposed in [42]. Unlike NAV, each DNAV is associated with a direction and a width, and multiple DNAVs can be set for a node. For each DNAV, there is a unique timer representing the channel states in that direction. Based on the DNAV fields, the node may be blocked from transmitting in some directions, but is allowed to transmit in other directions. This mechanism is known as directional virtual carrier sensing (DVCS) [42] and was proven to be a useful protocol optimization that reduces the exposed terminal problem and increases the spatial reuse of the channel significantly. Figure 4.2 illustrates the DNAV mechanism. Node A sets its DNAV for the beam towards node C to avoid interfering with the ongoing communication between node B and node C. Based on the concept of DVCS, node A is not allowed to transmit to node D but can freely communicate with node E.

In case of omnidirectional RTS/CTS, DNAV can be unnecessarily set for directions where the channel will not busy. In [13], the authors propose to include the beam indices to be used for the imminent directional communication in the omni-directional RTS/CTS. Based on this information, neighbors of the communicating nodes decide which beams should be blocked as they might cause interference. A similar situation may arise with the case of multi-directional sequential RTS/CTS. Taking deafness into consideration, another optimization is proposed in [16] called enhanced DNAV (EDNAV). The EDNAV scheme consists of a DNAV table and a deafness table (DT). Whenever a node has a packet to be sent over one direction,

Fig. 4.2 A scenario to illustrate the DNAV mechanism

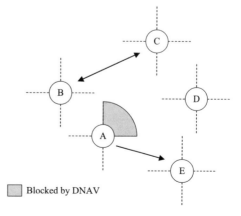

☐ Blocked by DNAV

both DNAV and DT are consulted. On the other hand, upon reception of a packet the node will either modify its DNAV or its DT, not both. If the node lies in the communication path between the transmitter and the receiver, the DNAV is to be modified. The DT is modified whenever the node receives either an RTS or CTS packet. In this case, the node is certain not to lie within the communication path of the oncoming transmission.

4.1.3 Idle Listening Mode

A node that is not receiving and does not have packets to send is refered to as an idle node. Intuitively, an idle node should reside in the omni-directional mode since it cannot anticipate the direction from which the signal might arrive. However, if the MAC protocol requires the idle node to reside in an omni-directional mode, the communication range is limited. Despite that limitation, this is the common design choice in directional MAC protocols. Another approach is to perform directional idle listening [28, 37]. This is done through continuous directional scanning to sense all directions sequentially. During this sweeping phase, if an activity is sensed in a certain direction, the rotating antenna beam stops and starts receiving the incoming signal. Hence, a source node who wishes to transmit needs to capture the rotating antenna beam prior to packet transmission. This could be done using an out-of-band tone which is long enough to capture an idle sweeping receiver. The advantages of employing directional idle listening are the full exploitation of the extended communication range and the alleviation of the hidden terminal problem due to asymmetry in gain. On the other side, it introduces initial delay, increases the problem of deafness and requires more power consumption.

4.1.4 Backoff

In part of CSMA/CA, a backoff mechanism is used for contention resolution. While remaining in the directional mode during the backoff period can prevent a node from getting unnecessarily captured by surrounding communications, the deafness problem is aggravated [8]. However, if each packet requires node B to carrier-sense, backoff, and communicate directionally, then multiple back-to-back packets would require a node to be in the directional mode for a long period of time which may eventually results in a severe deafness impact on one or more of its neighbors. Therefore, a design choice that has been widely adopted with directional MAC is to perform the backoff in an omni-directional mode.

Most directional MAC protocols have not altered the binary exponential backoff (BEB) mechanism of IEEE 802.11. Since BEB was originally designed with the implicit assumption that all node is the network are equipped with omni-directional antennas, it may not be suitable to be used in the presence of beamforming antennas.

This is mainly because BEB is an over-conservative mechanism which assumes all transmission failures are due to collisions. Hence, the use of BEB can limit the possible spatial reuse and aggravate the deafness problem. To address the limitations of BEB in the beamforming environment, more aggressive backoff mechanisms are proposed in which the contention window (CW) remains constant [32] instead of exponentially increase. This ensures the sender contends for the channel quickly in case deafness was the reason the receiver did not respond to a RTS message before. A more interesting idea is to adjust the boundaries of the CW based on the event that triggered the backoff [35]. For example, if the channel is busy, the node backs off with no change in the CW while if there is no CTS received, the CW is increased linearly and if there is no ACK received, the node implements an exponential increase of the contention window. When an ACK is received, the CW size is decreased exponentially.

One approach to optimize the traditional backoff mechanism for the case of beamforming antennas is to use separate backoff counter for each direction [28]. Using beamforming antennas gives the opportunity to adapt independently to the traffic conditions in different directions. For a node, one direction could be highly congested while the other directions might be less congested. Using the same back-off for all directions is an ill-fitted approach which has an adverse effect on the overall channel utilization.

A revamp in the design of the backoff process is introduced in [3]. Following a transmission failure due to either deafness or collision, the sender should halt the retransmission of the failed packet for a certain period of time as it happens in traditional backoff mechanisms. However, the sender is not forced to undergo idle backoff phase but can rather take the opportunity of transmitting other outstanding packets in other directions. Remaining idle during the backoff period is not suited for the case of beamforming antennas as it results in channel under-utilization. This novel non-idle backoff mechanism increases the spatial reuse and simultaneously reduces the impact of the deafness problem.

4.1.5 Beamforming Information

Although the neighbor discovery process does not lie within the domain of the MAC layer, it has a great impact on its operation. In the presence of beamforming antennas, neighbor discovery is not limited to identifying the nodes within the communication range but the beamforming information is also essential. The beamforming information is usually decided based on either the relative position of the nodes [24] or the Angle-of-Arrival (AoA) estimation [42]. The location-based beamforming requires additional hardware such as GPS and implicitly assumes that a Line-of-Sight (LOS) exists between the nodes. Since this assumption may not be true in multi-path environments, the AoA criterion is usually preferred. The AoA information is usually cached to be used for future transmissions.

A MAC protocol that employs directional RTS transmission requires the source node to be aware of the beamforming information of the intended destination. Some directional MAC protocols assume the beamforming information is provided by upper layers such as routing. Routing control packets such as Hello packets are broadcasted periodically and can be used to estimate the directions of the neighboring nodes. However, staleness of beamforming information could occur unless this information is collected on a per-packet basis.

Control messages at the MAC layer could be used to estimate the beamforming information. Overhearing ongoing RTS/CTS packet help the nodes to update the beamforming information of their neighbors. Moreover, explicit training sequences could be used to accurately estimate/update the beamforming information more accurately.

4.1.6 Tones

For adhoc networks with omni-directional antennas, busy tones can be used to resolve both the hidden and the exposed terminal problems. For example, in DBTMA [17], the RTS/CTS packets are used to turn on transmit and receive busy tones until the data transmission is completed. A node that hears any of the busy tones will defer.

For directional MAC, tones appear as a design choice to address the beamforming-related challenges. Tones are usually transmitted directionally to protect ongoing communication from collisions [28]. They can also be transmitted omni-directionally to announce the start and/or the end of a communication that may be a source of a deadness situation [8]. Moreover, if directional idle listening is employed, tones are used prior to packet transmission in order to capture the rotating antenna beam of the idle receiving node [46].

The drawback of using tones is the additional required hardware that increases cost and complexity. In addition, tones are commonly transmitted on a dedicated control channel which offsets the bandwidth.

4.1.7 Synchronized Access

Another design choice to handle the MAC challenges is to perform synchronized access rather than random access. By separating the transfer of control and data packets in time, the location-dependent carrier sensing problem are alleviated. Time is divided into frames with the data transfer sub-frame is preceded with a contention-based reservation sub-frame. Most synchronized directional MAC protocols [23, 47, 50] assume the availability of network-wide synchronization which is considered impractical specially in multi-hop wireless networks. Few recent synchronized MAC approaches [6, 41] are based on local synchronization. The first winning node-pair in the contention-based period decide the size of control and data sub-frames. Although this approach alleviates the complexity of global synchronization, setting the size of the control and data sub-frame is a critical tradeoff between under-utilization and poor spatial reuse. Moreover, communication between certain neighboring node-pair may not be properly scheduled if each of them lies in a different synchronized zone.

4.2 Overview of Directional MAC Protocols

The problem of designing an efficient MAC protocol for wireless ad hoc networks with beamforming antennas has been of a great interest during the last decade.

4.2.1 Classification

In this section, we present a taxonomy of the proposed directional MAC protocols as shown in Fig. 4.3. The MAC protocols can be broadly classified into random access protocols and synchronized access protocols. Random access protocols allow the stations to access the shared medium randomly through contention with each other. Synchronized access protocols allow the stations to access the medium based on a predetermined schedule which can be achieved through local and/or global synchronization.

A substantial number of directional MAC protocols presented in the literature belongs to the former category. Most random access protocols rely on the concept of Carrier Sensing Multiple Access (CSMA) in which physical carrier sensing is performed before initiating transmission. Random access protocols can be further classified into sub-categories according to the tool(s) used to handle MAC main challenges such as deafness and hidden terminals. The first sub-category of directional MAC protocols rely solely on control packets in particular RTS/CTS packets traditionally used for collision avoidance. The second sub-category employs busy tones that are usually transmitted on a dedicated control channel. The protocols that rely on the control packets can be further classified based on how the initial control packet (i.e. RTS packet) is transmitted. (1) Omni-directional RTS: The RTS packet is transmitted in all directions with the antenna operating in an omni-directional mode. (2) Uni-directional RTS: The RTS packet is transmitted directionally towards the direction of the intended destination only. (3) Multi-directional RTS: The RTS packet is transmitted towards some or all available directions. The multi-directional transmission could be either sequential or concurrent. If the antenna pattern is formed of a single beam, the multi-directional transmission could be achieved by transmitting copies of the packet sequentially over different directions (one direction at a time). When the beamforming antenna is capable of forming multi-beam antenna pattern, the packet could be transmitted to multiple directions concurrently at the same time.

Throughout the literature, we found some directional MAC protocols that belong to the category of synchronized access protocols. The basic idea is to coordinate conflict-free transmissions to occur simultaneously which requires some sort of synchronization between the nodes. Time is usually divided into frames and each frame consists of sub-frames which are simply a group of time slots. In one sub-frame, channel contention is usually used to perform a schedule for contention-free data transmission in the rest of the frame. Since achieving global synchronization is considered difficult in multi-hop wireless networks, recent protocols have chosen to rely on local coordination between neighboring nodes.

Aside from the above taxonomy, MAC protocols for wireless ad hoc networking with beamforming antennas can be classified in different ways. One classification could be according to the antenna capabilities whether switched-beam antennas, steered beam antennas or adaptive antennas with null capabilities. Another classification could be based on supported communication range which is limited by the antenna modes at each side of the wireless link. A third classification is whether the MAC protocol use a single channel or multiple channels. Directional MAC protocols can also be classified based on the power awareness of the protocol or its IEEE 802.11 compatibility. It is worthy to note that the above classes are not independent of each other and hence one directional MAC protocol may belong to more than one class. In this work, our classification is based on the taxonomy shown in Fig. 4.3

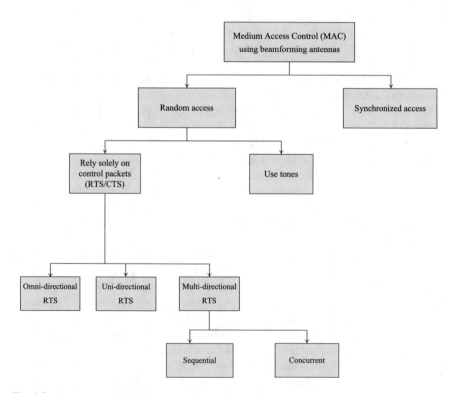

Fig. 4.3   A taxonomy of MAC protocols for wireless ad hoc networks with beamforming antennas

which provides the fine granularity needed to understand the benefits and tradeoffs associated with the surveyed protocols. In the next two sections, we will review the operation of thirty eight directional MAC protocols that best represent the progress in this field.

4.2.2 Random Access Protocols

Due to the lack of a pre-determined access schedule, stations compete to access the shared medium through random access. In case of conflict occurrence, a distributed conflict resolution algorithm is used to resolve it. Most random access protocols rely on the concept of CSMA. A station wishing to access the wireless medium performs carrier sensing before initiating transmission. If the medium is idle, the station is allowed to transmit. If the medium is sensed busy, the station defers transmission for a random period of time. In traditional wireless networks with omni-directional antennas, collision avoidance mechanisms have been widely used to improve the performance of CSMA-based protocols. Collision avoidance is performed using control packets (e.g. RTS/CTS packets) and/or busy tones. In the context of beamforming antennas, similar mechanisms are employed to address the major MAC challenges with beamforming antennas such as directional hidden terminals and deafness.

4.2.2.1   RTS/CTS-based Protocols

In this section, we overview the directional MAC protocols that rely on the control packets in their operation. These protocols are inspired by the operation of the IEEE 802.11 DCF [20] due to its simple design and its wide spread usage. As discussed in Sect. 3.1, the IEEE 802.11 DCF MAC is based on the concept of CSMA/CA. Small RTS/CTS control packets are exchanged prior to data transmission as part of the collision avoidance process.

Since the design of IEEE 802.11 MAC implicitly assumes an omni-directional antenna at the physical layer, researchers have looked into modifying its operation in order to exploit the potential benefits of beamforming antennas. A common design choice adopted by directional MAC protocols designers is the directional transmission of both the data and acknowledgment packets. However, there are several variations for how RTS/CTS packets are transmitted in order to deal with the challenges associated with beamforming antennas.

Protocols that Use Omni-Directional RTS

Nasipuri et al. in [33] are among the first to investigate appropriate MAC protocols for multi-hop ad hoc networks with multiple antennas. They assume a very simple antenna model in which each node is equipped with multiple directional antennas forming non-overlapping beams that can collectively cover the entire plane. In this protocol, the authors propose that the data and its acknowledgement should be exchanged directionally in order to reduce the interference, thereby increasing the network throughput. Since the neighbors' location information may not be available at each node, especially with frequent node movements, they propose to send both RTS and CTS omni-directionally (ORTS/OCTS). Idle nodes listen to the surrounding medium in an omni-directional mode. When a node receives RTS for itself, it marks the beam from which it received the packet and responds with the omni-directional CTS. Upon receiving the CTS in response, the sender node also knows the direction of the intended receiver by noting the antenna beam that received the RTS packet with the maximum power. Each neighboring node that receive either the RTS or CTS, begin an off-the-air period for the duration specified in the RTS/CTS packet similar to IEEE 802.11 NAV. Although the reported results show an increase in the total throughput, the proposed protocol has some limitations. Since the channel reservation is done in an omni-directional mode, the communication range is limited by the omni-directional gain. Also, the spatial reuse is severely affected by the need to transmit the RTS/CTS omni-directionally.

Fahmy and Todd propose in [14] the Selective CSMA with Cooperative Nulling (SCSMA/CN) protocol for ad hoc network stations with adaptive antenna arrays. They propose to transmit all the packets omni-directionally and exploit the nulling capabilities of the receiving antenna to dynamically null potential future interfering packet transmissions. After the exchange of RTS/CTS packets, the source node along with all of the nodes that received the CTS packet simultaneously transmit a short Cooperative Nulling (CN) packet so that the beamforming weights at the destination node are calculated. The beamforming antenna attempts to maximize the desired signal and null those interfering transmissions. Following this, the destination node and all of the neighbors of the source node send CN packets in the same fashion so beamforming can be performed at the source node. Using this method, the reception of DATA and ACK packets is protected. SCSMA/CN employs selective CSMA in the sense that carrier sensing is used only if the ongoing packets are unprotected RTS/CTS packets. The presented results show capacity improvements over IEEE 802.11 and protocols with steered-beam antennas (no nulling capabilities). However, the performance of the protocol is limited to the available degrees of freedom of the antenna array.

In [32], Mundarath et al. also consider ad hoc networks with adaptive antenna arrays. They propose NULLHOC MAC protocol that can work in multi-path environments. In the NULLHOC protocol, the total bandwidth is divided into two orthogonal channels: a Data Channel (DC) and a Control Channel (CC). The access rights to the DC are obtained through three control packets transmitted omni-directionally on the CC. The source sends RTS packet that contains the antenna weights the node will use for receiving the ACK. If the destination is able to involve in this communication, it responds with CTS packet that contains the receiving and transmitting antenna weights. Then, the source reserves the access right to the DC by sending a Data-Send (DS) control packet that contains the antenna weights that the node will use while transmitting the DATA packet. Nodes that overhear either RTS, CTS, and/or DS record the details of the corresponding communication. When the nodes finish their communication, they have to wait for a fixed duration before they are allowed to initiate a new communication. This is done because these nodes may not be aware of new ongoing communications that started while they were communicating. The simulation results show that NULLHOC protocol provides up to a factor of two increase in throughput relative to IEEE 802.11. However, the throughput gains tend to saturate as the number of antennas increase due to increased control overhead.

In order to alleviate some of the problems facing medium access in the presence of beamforming antennas, Arora et al. propose a Directional MAC with Power control (DMAP) in [1]. They assume switched beam directional antenna with constant gain in the main lobe. They use separate control and data channels to solve hidden terminal problem due to unheard RTS/CTS messages at the expense of additional sophisticated hardware. Idle nodes listen omni-directionally to the data and control channels. When a node has a packet to send, it first sends the RTS omni-directionally with a common fixed power. Upon reception, the intended receiver estimates the Angle-of-Arrival (AoA), calculates a power control factor and encapsulates it in the Directional CTS (DCTS) sent to the source node. The source node uses the power control factor to calculate the sufficient transmit power needed to transmit the data packet. The power of the DCTS is scaled by a power-scaling factor that ensures that every potential interferer listening omni-directionally can hear the DCTS. Transmission of DCTS from minor lobes of the receiver at scaled power would also prevent potential interferers located in other directions. The authors claim this may resolve deafness as well. The simulation results show that DMAP improves the network throughput and reduces the energy consumption at the same time.

Protocols that Use Uni-Directional RTS

In [24], Ko et al. are the first to propose modifications for IEEE 802.11 DCF for ad hoc networks with directional antennas. They assume packets can be transmitted directionally or omni-directionally but packet reception can be done omni-directionally only. They propose the D-MAC protocol in which RTS is sent directionally (DRTS) towards the intended receiver to avoid unnecessary waiting time if one of the other directions is blocked. The basic assumption here is that each node knows the location information of each of its neighbors by means of Global Positing System (GPS) and each node transmits based on the direction derived from the physical location information. To avoid collisions at the receiver, omni-directional CTS is sent followed by directional DATA and ACK exchange. The simulation results show performance improvement due to the increase in the number of concurrent transmissions in the network. The results reported with this simple D-MAC have motivated a lot of research in the area.

Takai et al. propose in [42] the concept of Directional Virtual Carrier Sensing (DVCS) for contention based MAC protocols to make effective use of directional antennas, while also providing interoperability with omni-directional antennas. Three primary capabilities are added to the original IEEE 802.11 for directional communication with DVCS. First, each node caches estimated AOAs from neighboring nodes when it hears any signal. Using the AOA cache, a source node can transmit DRTS without the need of additional hardware. Second, when a node receives an RTS from a neighbor, it adapts its beam pattern to maximize the received power and locks the pattern for the rest of the communication. Also, the source node locks its beam pattern after CTS reception. Beam locking prevents the nodes from being distracted by signals from other directions. The third and main capability to support DVCS is the use of directional NAV (DNAV). Each node maintains a DNAV table which can consist of multiple DNAVs each has its own direction, width and expiration time. If a node receives a packet from a certain direction, it needs to defer transmissions only in that direction in which other communication is in progress. DVCS determines that the channel is available for a specific direction when no DNAV covers that direction.

Choudhury et al. generalize the ideas in [24] and propose a directional version of IEEE 802.11 DCF MAC under the name of "Basic DMAC" in [10]. Basic DMAC is considered the benchmark for directional medium access control protocols. The authors assume that an upper layer is aware of the neighbors of a node and is capable of supplying the transceiver profiles required to communicate with each of these neighbors. The MAC layer receives these transceiver profiles along with the packet to be transmitted. In Basic DMAC, RTS/CTS/DATA/ACK are all transmitted directionally. An idle node listens to the channel omni-directionally but when it receives a signal, its antenna system is capable of determining the Direction-of-Arrival (DoA) of this incoming signal. The receiving node locks onto that signal and receives it. The physical carrier sensing and the backoff phase are performed while the antenna is in a directional mode. Moreover, Basic DMAC performs DVCS using DNAV tables similar to [42]. In the context of Basic DMAC, most of the MAC challenges with beamforming antennas have been identified. The authors evaluate the trade-offs associated with Basic DMAC. The results show that directional communication has the potential to improve the performance in terms of aggregate throughput and end-to-end delay. However, the performance mainly depends on the topology and flow pattern in the network. Random topologies with unaligned flows perform much better than aligned topologies since the spatial reuse can be exploited.

In order to exploit the benefit of highest communication range with beamforming antennas, communication should be possible between nodes that are neighbors only when both the transmitter and the receiver are in directional mode known as DD-neighbors. To support the full range extension, Choudhury et al. propose the MMAC protocol in [10]. The MMAC protocol aims to transmit the data packet over the longest possible hops. Since the idle nodes reside in omni-directional mode, they propose to propagate the RTS over multiple hops to inform the DD-neighbors to beamform towards the transmitter. In MMAC, the MAC layer receives a packet from an upper layer containing the DO-neighbor route to the next DD-neighbor. A special RTS packet contains the DO-neighbor route is transmitted to the next neighbor on that route. Nodes along that route forward the RTS according to the encapsulated route. The special RTS gets highest priority and is forwarded with a preceding backoff. Once the RTS is received by the DD-neighbor, CTS, DATA and ACK are transmitted over the single long hop. The simulation results show that MMAC outperforms Basic DMAC in terms of aggregate throughput. The limitations of this protocol include the long delay of RTS propagation and the risk of losing RTS over multiple hops. Also, the intermediate multi-hop paths for RTS propagation may not always be available.

Kolar et al. in [25] identify the HoL blocking problem associated with directional MAC protocols with beamforming antennas and FIFO queuing. The authors propose a new greedy queuing policy that can be implemented within the DMAC protocol. Based on the DNAV table, the authors propose using the least wait time to pick a packet for transmission. The simulation results show that the new queueing policy outperforms the existing one in terms of overall throughput and end-to-end delay. However, the proposed scheme does not consider the effect of deafness, which may cause the DNAV entries to be invalid.

In [35], Ramanathan et al. propose and implement a complete system for ad hoc networks with directional antennas called UDAAN. The UDAAN-MAC protocol has two features that differentiate it from previous approaches which are a new backoff mechanism and the integration of power control. The authors propose a new backoff algorithm (called forced idle) in which the duration and the window adjustment mechanism depend on the type of event causing the backoff, for example whether the event is busy channel, missing CTS, or missing ACK. If the channel is sensed busy, the contention window remains constant. If CTS is found missing, the value of the contention window is increased linearly. In case of the absence of an ACK, the increase of contention window is exponential. Upon receiving an ACK the value of the contention window decreases exponentially. On the other hand, the UDAAN-MAC protocol is a power-controlled MAC. The RTS is sent at the power indicated in the radio profile sent with direction by the forwarding layer. The RTS contains the transmitted power and the source node's current receiver threshold. Using this information the receiver can adjust the transmit power for the CTS packets. The DATA and ACK are power adjusted in a similar manner. The UDAAN-MAC protocol performs power-controlled DVCS. The DNAV table contains the duration, the direction and the allowed power. The last field indicates the power above which interference will occur in this direction. This direction may still be used to transmit if it is deemed that the intended transmission is sufficiently low power so as to not bother the busy nodes.

In [31], Takata et al. address the deafness problem by a Receiver-Initiated Directional MAC (RIDMAC) protocol. By default, the RIDMAC protocol is a sender-initiated DMAC in which all packets are transmitted directionally. If a transmitter noticed that there is another packet addressed to the same receiver in the head of its queue, it appends the size of the next packet to the header of the current data frame. Each node maintains a polling table and uses the information in the header of the data frame to update its table. After exchanging the DATA/ACK frames, the transmitter and the receiver check their own polling table whether potential deafness nodes exist or not. If more than one node is registered in the polling table, the least recently transmitting node is polled using a directional Ready-To-Receive (RTR) packet. Once RTR is received, the polled node, that was possibly suffering from deafness, transmits the data frame.

Choudhury and Vaidya propose a Capture-Aware Directional MAC (CADMAC) to address the MAC-layer capture problem in [9]. The CADMAC protocol aims to prevent a node susceptible to capture from operating in the omni-directional mode while idle. If the capture directions are known, the node forms a multi-beam pattern with main-lobes in directions other than the capture directions. CADMAC assumes time is divided into cycles with each cycle subdivided into ON and OFF durations. During the ON duration, the MAC layer records every received packet and the beam used to receive it. If a beam proves to be the receiver of only capture traffic, then the beam is black-listed. At the end of the ON duration, CADMAC decides to turn off all black-listed beams for the next OFF duration. In CADMAC, RTS/CTS/DATA/ACK are all transmitted directionally but the DVCS is modified to be capture-aware. When a node overhears an RTS or a CTS packet on a particular beam, CADMAC recommends the physical layer to turn off that beam for the proposed duration. The simulation results show improvements in throughput and end-to-end delay when compared to capture-unaware directional MAC protocols.

Bazan and Jaseemuddin propose an Opportunistic Directional Medium Access Control (OPDMAC) protocol in [3]. The OPDMAC protocol aims to grasp the transmission opportunities offered by beamforming antennas by eliminating the use of the over-conservative binary exponential backoff algorithm commonly used by most directional MAC protocols. In OPDMAC, the node is not forced to undergo idle backoff after a transmission failure but can rather take the opportunity of transmitting other outstanding packets in other directions. This novel mechanism minimizes the idle waiting time, increases the channel utilization, reduces the impact of the deafness and prevents the head-of-line blocking. After each successful transmission, the node is forced to remain idle for a random period of time called the Listening Period (LP) even if it has packets outstanding for transmission. During LP, the node listens in an omni-directional mode. The listening phase is needed to reduce the transmission failures due to deafness and to allow each node to update its channel state information. The simulation results show that OPDMAC outperforms other protocols in terms of throughput, delay and fairness.

In [38], Shihab et al. propose the Directional-to-Directional MAC (DtD-MAC) protocol that requires the beamforming antennas to operate in directional mode only. Instead of the omni-directional idle listening, DTD-MAC performs directional idle listening through continuous directional scanning to sense all directions. Using DtD-MAC, communication is possible with DD-neighbors and the hidden terminal problem due to asymmetry in gain is alleviated. However, the problem of deafness is aggravated and the probability of collision is increased. To address these issues, the sender transmits multiple DRTS packets towards the receiver (up to $2M$ DRTS where M is the number of beams) in order to capture the continuously scanning idle receiver. Moreover, DtD-MAC requires the carrier sensing to be greater than the DATA period to avoid collisions. The reported results show that the large control overhead and excessive delay limit the performance of the protocol when the number of beams increases.

Fakih et al. propose the BMAC protocol for ad hoc networks with adaptive antenna arrays in [15]. BMAC performs joint channel gathering and medium sharing. The channel acquisition is performed proactively through a periodic training sequence. When receiving this training sequence, the channel to the corresponding node is estimated and the channel coefficients and the node identifier are saved in a channel table. When there is data ready to be sent, the source node sends a Beamformed

RTS (BRTS) to maximize the power at the destinations and make nulls towards the potentially interfering neighbors. When receiving the BRTS, the destination node calculates the exceeded power for further transmitted power correction and then it sends an Omni-directional CTS (OCTS) packet containing this correction factor. The results show that BMAC offers higher throughput than the conventional DMAC in multi-path fading environment.

Protocols that Use Multi-directional Sequential RTS

Korakis et al. propose the Circular RTS MAC (CRM) protocol in [26] which is the first protocol to employ the multi-directional sequential transmission of the RTS packet. The rationale is to inform all the neighbors about the upcoming communication using directional transmissions only and hence the protocol is able to achieve communication range extension as well. In CRM, the directional RTS is transmitted consecutively in a circular way until it scans all the area around the transmitter. The transmitter does not need to know the direction of the receiver. The duration field of the RTS packet is decreased by the RTS transmission period, every time an RTS packet is transmitted in the cycle. The receiver replies with a directional CTS after the conclusion of the circular RTS. Although CRM addresses some of the challenges facing medium access with beamforming antennas, the control overhead of the protocol is significantly large.

In [22], Jakllari et al. propose the Circular RTS and CTS MAC (CRCM) protocol that requires circular RTS and circular CTS packets prior to data transmission. Similar to CRM [26], the sender transmits circular directional RTS packet to all directions and the receiver sends a directional CTS towards the sender. Different from CRM, CRCM requires the receiver to circularly transmit CTS to inform un-aware neighbors about the imminent communication. Unaware neighboring nodes are those nodes that are in the coverage range of the receiver but not in that of the transmitter. The CRCM protocol protects the ACK reception from collision and hence handles the hidden terminal problem due to the asymmetry in gain at the expense of additional delay and large control overhead.

Gossain et al. in [16] propose a MAC protocol for Directional Antennas (MDA) that also employs the circular directional RTS/CTS transmissions. A key difference from the previous protocols is that both the sender and the receiver transmit the circular RTS and CTS packets simultaneously after they successfully exchange a single directional RTS/CTS. This somehow decreases the delay and ensures the circular control packets are only transmitted after the original RTS is successfully received. To avoid coverage overlap of the circular RTS/CTS, MDA employs a Diametrically Opposite Directional (DOD) procedure. It is obvious that the MDA protocol needs a prior determination of the neighbors' location. This is performed using a directional neighbor table that is established during the route discovery process and maintained by overhearing packets at the MAC layer. In MDA, the overhead associated with the DOD RTS and CTS packets is optimized by sending these packets only through those directions where neighbors are found. Another new feature in MDA is the use of an Enhanced DNAV (EDNAV) mechanism that differentiate between collision avoidance and deafness avoidance. The EDNAV consists of two components: A

DNAV table which is modified when the node receives the first directional RTS/CTS packets and a Deafness Table (DT) that is modified when a node receives a DOD RTS/CTS. The simulation results show that MDA performs better than IEEE 802.11, Basic DMAC and CRM protocols.

Li and Safwat propose a DMAC protocol with Deafness Avoidance and Collision Avoidance (DMAC-DACA) in [30]. In this protocol, the basic directional RTS/CTS exchange is followed by sweeping RTS/CTS counterclockwise to inform all the neighbors about the upcoming communication. Deafness is avoided using a deafness neighbor table that use the sweeping RTS/CTS to record the deafness duration of neighboring nodes. The authors also address another type of deafness that occurs due to the MAC-layer capture problem discussed is Sect. 3.2.5. The location information, retrieved by GPS, is added to the RTS/CTS frames. Using this information, the node that receives RTS/CTS can update the record in its deafness neighbor table if any of the neighbors is in the coverage area of the upcoming transmission. The idea of allowing the reservation messages to carry information about the direction of transmission was first proposed in [13] to balance the tradeoff between spatial reuse and collision avoidance. The DMAC-DACA protocol performs collision avoidance through the DNAV mechanism. A node updates its DNAV if the transmitter or the receiver node is a DD-neighbor of this node.

In [43], Takata et al. propose a Directional MAC with Deafness Avoidance (DMAC/DA) to address the tradeoff between deafness avoidance using additional control frames and the excessive overhead associated with them. In DMAC/DA, Wait-To-Send (WTS) frames are transmitted by the transmitter and the receiver after the successful exchange of directional RTS and CTS similar to MDA [16]. However, WTS frames are transmitted only to the directions where potential transmitters are located in order to reduce the control overhead. The potential transmitter is selected either based on the history of previous communications or by means of explicit next packet notification if possible. The simulation results show that DMAC/DA outperforms circular directional MAC protocols, especially when the numbers of flows and beams are large.

In [7], Chin proposes the SpotMAC protocol that is based on the use of pencil (narrow) beams. Pencil beams provide high spatial reuse and constrain the hidden terminal problem to a linear topology. SpotMAC uses an additional inverted RTS/CTS exchange to overcome the hidden terminal problem. A node that wants to transmit to a downstream neighbor must first ask its upstream neighbor for permission using an RTS-req packet. The upstream neighbor blocks transmission in that direction and responds with a CTS-ACK packet. The sender node will then undergo DRTS/DCTS/DDATA/DACK dialog with its downstream neighbor. Finally, the node will send an ACK-ACK packet to unblock the upstream neighbor. The above mechanism is very conservative since the upstream neighbor may be deaf to the RTS-req packet and hence the communication towards the downstream neighbor is unnecessarily blocked. The author proposes to optimize SpotMAC by enabling the inverted RTS/CTS exchange only if there is persistence interference form upstream neighbor. The use of pencil beams increases the probability of deafness significantly. Whenever a failure is encountered, SpotMAC allows the sender to contend for the channel quickly by backoffing for a random period of time derived from a constant contention window. This reduces the effect of deafness. If the number of failures exceeds a threshold, the contention window is increased exponentially. The results show that pencil beams can achieve very high spatial reuse in non-deafness scenarios.

Protocols that Use Multi-directional Concurrent RTS

Among the early attempts to exploit the capabilities of beamforming antennas in adhoc networks, Bandyopadhyay et al. in [2] propose an adaptive MAC protocol for wireless ad hoc networks using a kind of adaptive antenna arrays known as ESPAR. The ESPAR antenna is capable of forming multiple directional beams as well as multiple nulls. Each node periodically collects its neighborhood information and forms an Angle-SINR Table (AST). The AST specifies the strength of radio connection from each node to its neighbors at different particular directions. Using these information, a Neighborhood-Link-State Table (NLST) at each node is formed to determine the best possible direction of communication with any of its neighbor. According to the proposed MAC protocol, idle nodes remain in a selective multi-directional listening with their nulls steered towards active communicating nodes. Also, RTS and CTS packets are sent selectively multi-directional to avoid interfering with known ongoing communications. Moreover, communicating nodes should steer nulls towards directions that are selectively ignored in the RTS/CTS transmission since nodes in those directions are not aware of this communication and may interfere with it.

In [5], Capone et al. propose a Power-Controlled Directional MAC (PCDMAC) protocol for wireless mesh networks with adaptive antennas. A novel feature in PCDMAC is the transmission of the RTS and CTS packets concurrently in multiple directions with a tunable power per direction that is adjusted to avoid interference with ongoing transmissions. This is done to inform the maximum number of neighbors of the new transmission. PCDMAC employs a DNAV that has an additional entry specifying the minimum power gain to reach an active node. After the successful exchange of RTS/CTS packets, the DATA and ACK packets are transmitted directionally with the minimum required power to reduce the interference and increase the spatial reuse. The simulation results show that both the throughput and fairness are improved using the PCDMAC protocol.

4.2.2.2 Tone-Based Protocols

In this section, we review the directional MAC protocols that use tones as part of their operation. A tone is a pure unmodulated sinusoidal wave transmitted at a particular frequency. Tones do not contain any information and hence do not need decoding but only need to be detected. In traditional ad hoc networks, tones (known as busy tones) are typically transmitted by busy nodes on separate dedicated channels (narrow bands) to inform all the nodes in their neighborhood about the ongoing transmission and hence protect them from collisions. The disadvantages of using tones are the bandwidth offset and the additional required hardware. In the context of wireless ad hoc networks with beamforming antennas, tone-based MAC protocols use tones together with RTS/CTS control packets to perform collision and/or deafness avoidance.

In [19], Huang et al. extends the idea of the Dual Busy Tone Multiple Access (DBTMA) [17] for the case of Directional Antennas (DBTMA/DA). In the proposed protocol, the channel is split into a data channel for data frames and a control channel for control frames with the two busy tones, transmit busy tone (BTt) and receive busy tone (BTr), are assigned two separate single frequencies in the control channel. When a node has data to send and it cannot sense BTr, the node transmits an omni-directional RTS since the receiver direction is not known. When the RTS is received and the receiver does not sense BTt, it responds with a directional CTS and turns on the directional BTr. Upon receiving the CTS, the source nodes transmits the data frame directionally and turn on the directional BTt until the data transmission is completed. The simulation results show that the network performance is improved by applying directional antennas to DBTMA and the performance is also better than that of the IEEE 802.11.

Singh and Singh propose Smart 802.11 protocol for ad hoc networks with adaptive antenna systems in [40]. When a node has a packet to send, it beamforms towards the intended receiver and transmits a short sender-tone to initiate communication. All idle nodes that receive the sender-tone beamform towards the sender and enter a random defer phase before transmitting the receiver-tone. When the sender receives the receiver-tone, it transmits its packet and waits for the receipt of an ACK. If there is no ACK, it enters backoff as in IEEE 802.11. Since the proposed protocol does not take care of hidden terminals, the authors rely on dynamically forming nulls towards interferes as well as the use of forward error correcting codes.

Choudhury and Vaidya propose ToneDMAC in [8] which specifically addresses the problem of deafness. In ToneDMAC, the backoff phase is performed in an omni-directional mode to alleviate the possibility of deadlocks and prolonged periods of deafness. ToneDMAC uses a tone-based notification mechanism that allows the neighbors of a node to distinguish congestion from deafness and react appropriately. After the data communication is over, both the sender and the receiver transmit out-of-band tones omni-directionally to inform their neighbors about the end of their deafness period. The neighboring node, that detects a tone, can identify the originator using the frequency and the duration of that tone. If the tone-receiving node is in a backoff phase waiting to communicate with a tone-originating node, it preempts its long backoff phase, initializes its contention window, and backs off with the minimum contention window. This reduces the unnecessary waiting time induced by using exponential backoff following transmission failures caused by deafness. The simulation results show that ToneDMAC is effective in mitigating the adverse effects of deafness.

In [36], RamMohan et al. address the problem of hidden terminals due to unheard RTS/CTS. They propose Fragmentation-based Directional MAC with TONE (F-DMAC-TONE) protocol that does not assume separate data and control channels. F-DMAC-TONE uses a combination of three features to solve the problem. When a node returns from directional to omni-directional mode, it undergoes a pause period before attempting transmission in another direction. This pause period increases the probability that the node learns of the true status of the channel. Ideally, the pause period must be long enough for an ongoing communication to finish. However, such a long pause period will lead to wasted resources if there was no ongoing transmission resulting in increased delay and degraded performance. To address this issue, a second feature in F-DMAC-TONE is the fragmentation of packets into smaller chunks transmitted individually but acknowledged collectively. The third feature is the use of a short TONE signal in between fragments to inform other nodes capable of causing collisions with the ongoing transmission. The simulation results show a significant decrease in the number of collisions due to the unheard RTS/CTS problem. However, a marginal improvement in the throughput and delay performance is achieved. This is mainly because the hidden terminal problem is not that critical when compared to the deafness problem.

Kulkarni and Rosenberg in [28] propose the Directional Busy Signal Multiple Access (DBSMA) protocol that relies on the use of busy tones. In DBSMA, all the transmissions, receptions, and idle listening are performed directionally to achieve better connectivity. When a node is in an idle state, its directional antenna sweeps continuously to cover the whole region. When a node wants to transmit, the node transmits an out-of-band invitation signal which is long enough to capture an idle sweeping receiver. The invitation signal is followed by an RTS packet. The invitation signal locks sweeping antennas in one direction to receive the RTS and the intended receiver responds with a CTS packet. While in the reception mode, the receiver continuously transmits a busy signal to alleviate any possibility of collision from the hidden terminals. In DBSMA, if the sender senses a busy signal or busy channel in one direction, it may choose to communicate with another node in another direction. Moreover, DBSMA uses a separate backoff counter for each direction in order to adapt independently to the traffic conditions in different directions. The results show a performance improvement when compared to the CRM protocol [26]. However, the deafness problem is not addressed even though it is more severe with the directional idle listening proposed in DBSMA.

In [29], Li et al. propose the Flip-Flop Tone directional MAC (FFT-DMAC) protocol that utilize two pairs of tones to solve the deafness and hidden terminal problems. The first pair of tones are transmitted omni-directionally to announce the start and end of a communication, therefore, overcoming the deafness problem. The second pair of tones are sent directionally by the receiver towards the sender to solve the hidden terminal problem and to acknowledge the receipt of both RTS and DATA packets. In FFT-DMAC, each node maintains a "deafness nodes" list and "ongoing transmission nodes" list that are updated with the reception of tones. The simulation results show that FFT-DMAC outperforms Tone-DMAC in the number of successful packets received per second.

Dai et al. propose the Busy Tone Directional MAC (BT-DMAC) protocol for wireless ad hoc networks using directional antennas in [12]. BT-DMAC combines the use of two busy tones with the DNAV table [42] to solve the deafness and hidden terminal problems. When the transmission is in progress, the transmitter and the receiver turn on the transmitting busy tone BTt and the receiving busy tone BTr, respectively. Each tone is transmitted omni-directionally and is pulse-modulated with the node ID and the beam used for communicating. Any node hearing the busy tones learns the node IDs and the beam numbers from the tones and deduces whether the potential sending will interfere with the current transmission. The mechanism adopted by BT-DMAC increases the probability of successful data transmission.

In [46], Takatsuka et al. propose a Directional MAC protocol with Power Control and Directional Receiving (DMAC-PCDR) that mitigates the interference caused by directional hidden terminals and minor side lobes. The DMAC-PCDR protocol is based on the ideas proposed in [44, 45] but is implemented with less control overhead. DMAC-PCDR employs directional idle receiving through the continuous rotation of the antenna beam while the node is idle. Directional receiving eliminates the hidden terminal problem due to asymmetry in gain and the interference caused by the reception through the side lobes. In order to enable an idle receiver to receive the signal, each control packet (RTS or CTS) is transmitted with a preceding tone that is long enough for an idle node to hear it. The node which receives the preceding tone stops the rotation and receives the packet. On the other hand, DMAC-PCDR improves spatial reuse of the wireless channel and extends the communication range through transmission power control. It has three access modes and each mode is selected depending on the information available about the receiver's location.

4.2.3   Synchronized Access Protocols

Most of the challenges facing the medium access are related to the location-dependent carrier sensing adopted by random access protocols. An alternative approach to address these issues to better exploit the benefits of beamforming antennas is the use of synchronized access protocols. Based on the availability of synchronization among competing nodes, conflict-free data transmissions occur according to a pre-determined time schedule. To build feasible schedules, nodes exchange control packets in a contention-based phase prior to the data transmission phase. Other mechanisms could also be done before data transmission including neighbor discovery and accurate beamforming. Synchronization could be performed network-wide or local based on the protocol requirements.

In [39], Singh and Singh propose the DOA-MAC protocol for nodes equipped with adaptive antenna array in ad hoc network. DOA-MAC is based on the slotted ALOHA with each slot broken into three minislots. In the first minislot, all transmitters transmit a simple tone towards their intended receivers. The receivers then run a DOA algorithm to identify the direction of the transmitters. Each receiver forms its directed beam towards the direction that has the maximum power and forms nulls in all the other identified directions. The second minislot is the packet transmission minislot. After receiving the packet, the receiver rejects the packet if it is not the intended destination. Otherwise, the receiver responds with an ACK in the last minislot. The simulation results show that DOA-MAC achieves higher throughput than the Basic DMAC [10].

Zhang proposes a TDMA-based directional MAC protocol called LiSL/d in [50] and evaluates its performance in [51]. The LiSL/d protocol performs link scheduling through pure directional transmission and reception. Time is divided into frames and each frame is divided into three sub-frames. The first sub-frame is devoted for neighbor discovery which is performed through scanning and three-way handshakes. During the neighbor discovery process, the two nodes detect each other and agree on a future time slot at which the two nodes would reassure the connection and see if they can make any reservations. Reassurance and reservation are made at the second sub-frame when the two nodes point towards each other with their beams and exchange another three-way handshakes. The third sub-frame is for data transmission. The simulation results show that the LiSL/d significantly outperforms DVCS [42] and IEEE 802.11 when jamming is present.

Wang et al. in [47] propose a directional MAC protocol termed SYN-DMAC for ad hoc networks with synchronization. The timing structure of SYN-DMAC consists of three time phases in each cycle which are: Random access, DATA and ACK phases. The random access phase serves as channel contention for data transmission. Multiple RTS/CTS packets are exchanged and multiple data transmissions can be scheduled. The later scheduled data transmissions should not collide with previous scheduled transmissions. Upon receiving the directional RTS, the receiver replies with directional CTS if it can engage in the communication session, or with directional negative-CTS if it has been already committed to another session or the beam towards the sender is blocked. Upon receiving the CTS, the intended sender sends a directional CRTS (confirmed RTS) to confirm the reservation. In the DATA phase, parallel contention-free data transmission is achieved and in the ACK phase parallel contention-free ACK packets are sent.

In [23], Jakllari et al. propose a synchronous Polling-based MAC (PMAC) protocol for mobile ad hoc netwoks with directional antennas. In this protocol, the time is divided into contiguous frames and each frame is divided into three segments: search, polling and data transfer. In the search segment, each node searches for new neighbors by transmitting or receiving pilot tones directionally. If two nodes discover each other, they exchange a list of the available slots in their corresponding polling segments. Once a pair of nodes agree upon a polling slot, they communicate in the same slot periodically, frame after frame, until they lose connectivity. The polling slot allows the nodes to schedule data transfers in the third segment of the frame and also allows them to keep track of the direction of each other that may change due to mobility. The communication in the polling slot is preceded by the exchange of control packets to avoid collisions. In the data transfer segment, multiple data transfers take place according to the schedules formed in the polling segment. In PMAC, RTS and CTS messages are used prior to the data transfer in order to detect possible rare collisions. The results show that PMAC achieves high channel utilization even in mobile scenarios.

In [41], Subramanian and Das propose the Contention Window Directional MAC (CW-DMAC) protocol to address the deafness and hidden terminal problems using single channel and single radio interface. The idea is to separate the transmission of control and data packets in time without the need of network-wide synchronization.

Through contention, several RTS/CTS packets are exchanged omni-directionally within a control window duration. The size of the control window is defined by the sender of the first RTS/CTS packet. In CW-DMAC, the omni-directional RTS/CTS packets are overloaded with the beam index in which the actual DATA/ACK transmission will happen directionally. This information will help any other node in the same neighborhood to exchange RTS/CTS within the same control window if they do not interfere with the previously reserved transmissions. When a node receives an RTS but cannot send the CTS due to beam blockage, it instead sends a Negative CTS (NCTS) to inform the sender that the data transmission cannot happen without interfering with the already reserved transmissions. Upon receiving the NCTS, the sender sends a TC (Transmission Cancel) packet omni-directionally to inform neighbors that the current transmission has been canceled. At the end of the control window, the directional DATA packets are transmitted simultaneously followed by concurrent transmission of ACK packets. The simulation results show that CW-DMAC improves the network throughput when compared to Basic DMAC.

Wang et al. in [48] propose the Coordinated DMAC (CDMAC) protocol that also requires local synchronization only. The timing structure of CDMAC consists of a contention-period in which control packets are exchanged followed by two contention-free periods for parallel DATA and ACK transmissions. Different from CW-DMAC, CDMAC use three control packets (RTS/CTS/confirmed-RTS) for channel reservation, all transmitted omni-directionally. CDMAC does not require the neighbor directions to be known a priori. The beam indices to be used to transmit DATA/ACK packets are included in the CTS and confirmed-RTS packets. The master node-pair, those who first win the channel contention, specify the duration of the contention and contention-free periods. With the contention-period, multiple data transmissions can be scheduled as long as the new reservations take into consideration the previous ones. In addition to the beam blocking, the CDMAC protocol considers interference caused by side lobes. In CDMAC, the frame formats of both RTS/CTS resemble the IEEE 802.11 frames with DMAC extension to ensure compatibility. The simulation results show that CDMAC outperforms IEEE 802.11 and the Basic DMAC protocol.

In [6], Chang et al. propose Reservation Directional MAC (RDMAC) for multi-hop wireless networks with directional antennas. The RDMAC protocol operates in sessions with each session comprising a reservation period and a transmission period. In the reservation period, the first node to transmit the RTS defines the start and end time of the transmission period. Each node-pair exchanges four control packets. First, omni-directional RTS/CTS packets are exchanged so the node-pair can discover the beams to be used for directional transmission. The neighboring nodes that receive the ORTS/OCTS packets, estimate the direction of arrival and point their antennas towards the sender/receiver to receive the remaining control packets. The reserving nodes transmit directional RTS/CTS packets so the neighbor nodes can update their DNAV taking into consideration any possible interference caused by minor lobes. Similar to [41, 48], the reserving nodes must avoid initiating a transmission if this transmission conflicts with an already scheduled transmission. However, in RDMAC, if the destination of the head-of-line frame is busy in this session, the transmitter will search for a frame destined to the next non-busy receiver in the queue and hence avoiding the head-of-line blocking problem. The simulation results show that RDMAC outperforms CRM [26] in terms of throughput and delay.

4.3 Basic Directional MAC (DMAC)

The Basic DMAC protocol is presented in [10] to generalize and merge the ideas and optimizations proposed earlier to adapt IEEE 802.11 for beamforming antennas. Basic DMAC assumes an upper layer is aware of the neighbors of a node, and is capable of supplying the transceiver profiles required to communicate to each of these neighbors.

Channel reservation in Basic DMAC is performed using a RTS/CTS handshake, both being transmitted directionally. Considering the example in Fig. 4.4, node A wants to communicate with node B. The MAC layer at node A receives a packet from its upper layers along with the direction of the destination B, namely $beam^B$. Basic DMAC requests the physical layer to beamform towards node B and performs physical carrier sensing using $beam^B$. If the channel is sensed idle, Basic DMAC checks its DNAV table to find out whether it must defer transmitting in the intended direction. If it is clear to transmit in this direction, node A enters the backoff phase with the antenna pointing towards node B. When the backoff counter counts down to zero, node A transmits RTS packet to node B using beam $beam^B$.

In Basic DMAC, an idle node listens to the channel while it is in the omni-directional mode. When it receives a signal arriving from a particular direction, the antenna system is capable of determining the direction of arrival (DoA) of this incoming signal. The receiving node locks on to that signal and receives it. When the node beamforms in the receiving direction, it can avoid the interference from other directions. In our example, node B is able to receive the RTS from node A using $beam^A$. If the DNAV table at node A is not blocking transmissions on $beam^A$, the channel is sensed for SIFS period before CTS is transmitted to node A.

Neighboring nodes, that also receive the RTS, recognize that the RTS is targeted to another node (node B). Before discarding the packet, each node update its DNAV table with the captured DoA for the duration specified in the RTS packet. This prevents any neighboring node from transmitting any signal towards node A for the duration of the imminent communication. Similarly, all nodes overhearing the CTS packet update their respective DNAV tables to block any transmissions towards node B. For example, node C must defer transmissions on beams $beam^A$ and $beam^B$ but is allowed in other directions. Hence, communication between C and D can occur simultaneously with communication between A and B

Meanwhile, node A waits for the CTS using the same beam $beam^B$ that it had used to send the RTS packet. If it receives the CTS packet successfully, it initiates the transmission of the DATA packets using $beam^B$. Upon receiving the DATA packet, node B responds with the ACK packet. In case the CTS does not come back within a CTS-timeout duration, node A schedules a transmission and enter into the backoff phase with the contention window size is doubled as in IEEE 802.11. During the backoff phase, node *A* stays beamformed towards node *B*.

4.3.1 DMAC with Omni-Directional Backoff

The operation of Basic DMAC may result in severe cases of deafness. If the node has multiple back-to-back packets, it will remain beamformed for a long period of time and its neighbors may suffer from the impact of deafness. In the case of multi-hop wireless networks, if each node backoffs directionally, deafness chains can be formed and possibly may end up in deadlocks. Refer to Fig. 4.4, assume node *A* has multiple back-to-back packets to send to node *B* and node *E* has a packet destined to node *A*. In this case, node *A* appears deaf to node *E*, and node *E* is forced to backoff and wait. By performing backoff in a directional mode, node *E* will be deaf for another node's transmission (say node *F*) and so on.

A simple modification to Basic DMAC that can reasonably address this problem is that nodes have to switch back to the omnidirectional mode while counting down their backoff values [10]. Once the backoff timer expires, a node beamforms toward its intended receiver and initiates transmission. While backing off in the omnidirectional mode, a node senses the carrier busy only if a signal arrives from the direction in

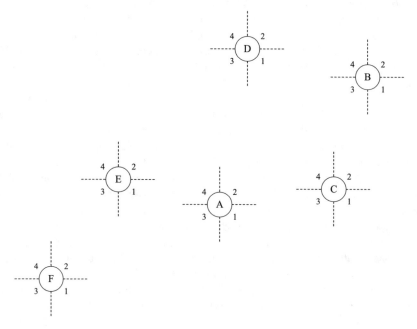

Fig. 4.4 An example topology for DMAC illustration

which the node intends to transmit. However, if a packet arrives from other directions, a node will be capable of receiving them. This mitigates the deadlock scenario. We denote the MAC protocol with this variation as DMAC-OM-BO.

4.4 Circular Directional RTS MAC (CDR)

The Circular Directional RTS (CDR) MAC [27] attempts to exploit directional antennas in wireless adhoc networks. The basic idea is that the RTS is transmitted directional sequentially, in a circular way, until it scans all the area around the transmitter. Unlike Basic DMAC, the transmitter does not need to know the direction of the receiver prior to initiating the communication. The transmitter starts transmitting its RTS in a predefined direction. Short afterwards it switches its antenna to the next beam and start sending the RTS in the second direction. It continues this procedure again and again until the transmission of RTS covers all the area around the transmitter. The RTS contains the duration needed for the conclusion of the whole four way handshake (RTS/CTS/DATA/ACK). Since the RTS packet is transmitted on all beams, the RTS transmission time depends on how many beams are remaining in the circular RTS transmission. For every RTS, the duration field is updated to reflect the remaining duration. In CDR, if the DNAV does not allow the transmission in a certain direction, the node cannot sent RTS in this direction and enters a silent mode for a duration equivalent to an RTS packet transmission. In this way, the transmitter respects the on going transmission without jeopardizing the potential of increasing the spatial reuse of the channel. When the transmitter completes the circular directional transmission of RTS, it listens to the medium in an omnidirectional mode to receive the CTS.

The receiver that hears the RTS packet addressed to itself will respond back with a CTS packet. However, it must wait until the end of the circular RTS transmission and afterwards it sends a directional CTS towards the direction of the transmitter of the RTS. Using selection diversity, the receiver recognizes the beam by which it receives the signal and so it will transmit the CTS using the same beam. Moreover, the receiver can use the beam information to estimate the wait time before it initiates the CTS transmission. Consider the topology in Fig. 4.4, if node A has a packet to send to node B, node A transmits circular directional RTS starting with beam 1 for example. When node B receives the RTS using beam 3, it sends the CTS back using beam 3. In each RTS, the transmitter sends, in addition to other information, a number which indicates the beam by which the packet is transmitted. In our example, node A informs node B that the RTS was transmitted using beam 2. Given that the antenna at each node has four beams, node B can deduce that node A still has to transmits two more RTS packets. Node B defers CTS transmission for the transmission time for two RTS packets in addition to SIFS period. During this deferral time, node B is locked ignoring the reception of any other packets. Upon receiving the CTS, the transmitter recognizes that the packet is received through beam 2 and uses it to transmit the DATA packet and receive the ACK packets.

Table 4.1 Location table at node E

Neighbor	Beam	Neighbor's beam
A	1	4
B	2	3
C	1	4
D	2	3
F	3	2

In CDR MAC, each node maintains a location table with one record for every neighbor it heard from. Initially the table is empty but it is updated in every reception. In every record, the node maintains the following information: The neighbor, the beam by which the node communicates with this neighbor and the neighbors beam by which it communicates with this node. In this way, every node maintains pairs of beams that are used for a direct transmission with its neighbors. Refereing to Fig. 4.4, the location table at node E is shown in Table 4.1. It is worthy to note that, as mentioned before, the location information is not used to facilitate the four way handshake communication due to the circular directional RTS transmission. Nevertheless, this information is useful for DNAV mechanism described below.

Due to the circular directional RTS transmission, the neighbors are informed about the intended transmission since the RTS contains the duration of the whole four way handshake. The CDR MAC protocol uses a DNAV table to keep track of the directions and the corresponding durations towards which the node must defer. However, it used a DNAV update scheme that is different from the one used in Basic DMAC. Contrary to Basic DMAC and similar protocols, a node using CDR MAC does not update the DNAV table based on the direction of reception of the RTS packet. Instead, the CDR MAC uses the location table along with specific information included in the RTS packet. In addition to other information, the transmitter of the RTS packet includes the pair of beams to be used in the DATA/ACK exchange by itself and the intended receiver. Every neighbor that receives the RTS examines its own location table to find the beams through which it use to communicate with the node-pair of the intended communication. If one of these beams coincides with the respective beam of the RTS packet, then the neighbor defers the transmission using this beam for the specified duration. Using the same example of node A communicating with node B in Fig. 4.4, the transmitted RTS packet contains the tuple $<A, B, 2, 3>$ which means node A will use beam 2 and node B will use beam 3 for their intended directional communications. Node E for example will check its location table and recognize that node A can hear it using beam 4 and node B can hear it using beam 3. Therefore, it realizes that it can interfere only with the reception at node B and so it updates its DNAV table to defer transmission through beam 2. This mechanism can handle directional hidden terminals problem more efficiently.

4.5 Opportunistic Directional MAC (OPMAC)

4.5.1 Main Design Considerations

In the case of omni-directional antennas, collision is the major reason for transmission failures. Hence, the binary exponential backoff algorithm is needed for contention resolution. Moreover, since a transmission reaches all the receivers in the sender neighborhood, the idle backoff phase is mandatory when a transmission failure occurs. On the other hand, when beamforming antennas are used, the channel is spatially divided and a transmission in one direction is not sensed in other directions. This major benefit of beamforming antennas should be fully utilized. A missing acknowledgment indicates a transmission failure that could be due to collision or deafness. In either case, the receiver is not currently ready to receive the packet and the sender should halt the packet retransmission for a certain period of time, as it happens in IEEE 802.11 backoff process that is typically employed by all directional MAC protocols. However, the directional MAC protocol should not force the sender to remain idle during this backoff period as implicitly assumed by the existing protocols. Remaining idle during the backoff period introduces unnecessary blocking time that results in channel underutilization and a significant increase in the delay. Instead, during the period the node is forced to backoff from transmitting in one direction as a result of transmission failure, it can take the opportunity of attempting transmission of other outstanding packets in other directions. In other words, the need to backoff for a random period of time before retransmission in one direction should not block packet transmissions in other directions. This active backoff procedure helps in enhancing the spatial reusability of the wireless channel to a great extent.

Observation #1: *In case of a missing acknowledgment, the node should not be forced to remain idle between the retransmission attempts as long as it has other packets to transmit in other directions.*

In IEEE 802.11 MAC, each node should go into an idle backoff phase after each successful transmission to ensure that backlogged nodes do not take control of the medium for long periods of time. This seems to be unnecessary in the case of beamforming antennas. The time the node spends in transmitting a packet in one direction can serve as a backoff duration for channel contention in another direction. Hence, contention resolution can be achieved by avoiding successive packet transmission in one direction rather than forcing the node to remain in an idle state. However, a critical deafness problem could arise. If a node has a backlog of packets to different neighbors residing in few directions, it would appear deaf to all the nodes in the other directions. This problem is similar to the original deafness problem in the context of DMAC [8]. The commonly adopted solution by most existing directional MAC protocols is to perform the backoff phase in an omni-directional idle state, which severely reduces the spatial reuse. The idle backoff, though seems beneficial, is prohibitive of taking advantage of observation #1, therefore, we suggest that alleviating the deafness chains should be decoupled from the backoff algorithm. To alleviate the persistent deafness, each node should regularly listen to the medium omni-directionally. Although this listening phase resembles the IEEE 802.11 back-off phase, its rationale and overhead are substantially different. The listening phase is needed with beamforming antennas to reduce the transmission failures due to deafness and to allow each node to update its channel state information, which is significantly different from the idle backoff algorithm that is originally designed for contention resolution. Moreover, the frequency of its occurrence is also different since it is not essential to enter the listening phase after each transmission failure as in the case of the IEEE 802.11 backoff phase.

Observation #2: Each node should regularly visit an omni-directional idle state to prevent persistent deafness.

Observation #3: In the case of beamforming antennas, the backoff phase and the listening phase should be decoupled.

The deafness problem is the most critical challenge facing multi-hop wireless networks with beamforming antennas. Although deafness occurs as a result of a transmission failure when the receiver is beamformed towards another direction, the sender's reaction escalates the problem. Upon detection of the failure, the binary exponential backoff algorithm is invoked resulting in channel underutilization, degradation in the network capacity, increase in the packet drops and unfairness in channel access. Most of the solutions proposed in the literature focus on reducing the occurrence of deafness by informing neighboring nodes about the upcoming transmission that may lead to deafness. This includes either omni-directional RTS/CTS transmission [33] or sequential directional RTS/CTS transmission over beams other than the receiver's direction [16, 22, 27, 43]. Although these approaches may reduce the occurrence of deafness, they cannot completely eliminate it as the overhead control packets may suffer from deafness themselves in addition to possible collisions. However, the main drawback of these techniques is the additional overhead that reduces the network capacity and increases the delay [43], which essentially offsets the benefits of spatial reuse that they try to exploit. Therefore, we need to address the deafness problem with either no or substantially reduced additional overhead so that the gain due to spatial reuse is not offset. Without the need of a deafness notification, we suggest the node that detects a packet failure should react in a way that alleviates the negative impact of deafness. The ideal behavior should minimize the blocking time, avoid the channel underutilization, reduce the correlation between the retransmissions, and avoid involving in an unfair backoff. The rationale behind our approach of having each node relieves the deafness problem on its own is the fact that deafness has no harmful impact on any other ongoing communication. In contrary, the hidden terminal problem may become more destructive by harming the ongoing transmission; therefore, a node is required to inform the neighborhood a priori to protect its transmission.

Observation # 4: Each node should react to transmission failures in a way that mitigates the impact of deafness.

Observation #5: *To leverage the benefit of spatial reuse, overhead should be minimized.*

The use of beamforming antennas introduce new hidden terminal problems in which the regular RTS/CTS fails to inform the hidden nodes about the ongoing communication. The hidden terminal problem due to asymmetry in gain is shown to be very rare [11], while collisions due to unheard RTS/CTS can occur more frequently. In this case, the virtual carrier sensing fails because it was performed while some nodes are beamformed towards other directions. This results in a loss of the channel state information whenever the node is beamformed. To reduce the effect of the hidden terminal problem, the node should try to retrieve its channel state information before each transmission attempt.

Observation #6: *To reduce the hidden terminal problem due to unheard RTS/CTS, the node should try to retrieve its channel state information before transmission.*

Considering no additional QoS requirement, the FIFO queuing policy works fine in the case of omni-directional antennas since all outstanding packets use the same medium. If the medium is busy, no packet can be transmitted. However, in case of beamforming antennas, FIFO leads to the HoL blocking problem [25]. In order to improve the spatial reuse, the packet scheduling policy should not block the transmission of any ready packet.

Observation #7: *The packet scheduling policy should enable the transmission of any ready packet, thus eliminates the HoL blocking.*

4.5.2 Protocol Description

Based on the set of observations presented in the previous section, we propose an opportunistic directional MAC protocol called OPDMAC for multi-hop wireless networks with beamforming antennas. The OPDMAC is a contention-based directional MAC protocol that aims to maximally harness the benefits of spatial reuse by minimizing the idle waiting time and exploring the new transmission opportunities. Although not mandatory, OPDMAC employs RTS/CTS exchange before data transmission. All messages are sent directionally while the idle node listens to the medium in an omni-directional mode. The OPDMAC uses the DNAV mechanism [42] for the directional virtual carrier sensing. We assume that an upper layer (e.g. routing layer) is capable of providing OPDMAC with neighbors' directions. This assumption is common among various directional MAC protocols [8, 10, 16, 24, 31, 43], which is justified because routing protocols usually learn the direction of neighbors through the reception of control packets such as route request and route reply packets during route discovery or periodic HELLO packets.

4.5.2.1   RTS Transmission

When a node ends its listening period, which is described in Sect. 4.5.2.5, it scans the packets in its non-empty link layer queue sequentially in the order of their arrival time to pick the first unblocked packet for transmission. The node attempts to transmit the packet by beamforming in the direction of the intended receiver and starting directional carrier sensing. If the medium is sensed idle for a DIFS period, the node transmits RTS packet. If the medium is sensed busy during the carrier sensing, the node has to defer transmission on this beam, however, it can still transmit over other beams. Accordingly, OPDMAC allows the node to rescan its queue and chooses the next unblocked packet and attempts transmitting it. The node beamforms in the new direction and starts the carrier sensing again. When the node succeeds in sending RTS, it initiates a wait-for-CTS timer and remains beamformed in the same direction.

4.5.2.2   RTS Reception and CTS Transmission

When a node receives RTS intended for the node itself, it performs a directional carrier sensing for a SIFS period. If the medium is sensed idle, it sends CTS in response. All other nodes that receive RTS not destined to them, update their DNAV table.

4.5.2.3   CTS Reception and DATA/ACK Exchange

Similar to most directional MAC protocols, when the sender receives CTS within the CTS-timeout duration, it sends the DATA packet after SIFS period. Upon receiving DATA, the receiver responds with the ACK packet indicating successful reception of the DATA packet. All other nodes that hear CTS, DATA and ACK packets, update their DNAV tables accordingly.

4.5.2.4   Missing CTS

If the sender does not receive CTS within the CTS-timeout duration, this means that the receiver is not currently ready for receiving the DATA packet. Since the sender could not distinguish between deafness and collision, it should not continue contending for the channel in this direction for a certain period of time similar to the backoff process of IEEE 802.11 generally employed by all directional MAC protocols. But, instead of forcing the sender to remain idle potentially going through exponential rounds of backoff, the OPDMAC allows the node to recheck its queue and try sending another packet in a different unblocked direction. *The period the node spends to transmit a packet in the second direction serves as a backoff period for the first direction.* This novel mechanism allows the node to be active during the backoff state and hence is able to minimize the delay and enhance the spatial reuse

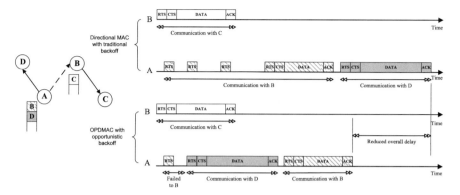

Fig. 4.5 A scenario illustrating the benefits of the opportunistic backoff employed by OPDMAC

significantly. Figure 4.5 shows an illustrative example. When node A fails to communicate with node B, it opportunistically replaces the traditional idle backoff time by a useful transmission of the packet destined to node D. After node A completes its transmission attempt to node D, it retransmits the packet destined to node B.

Occasionally, the node may not find any unblocked packet so it is forced to enter an idle backoff state. In OPDMAC, the node backs off in an omni-directional mode for a random time derived from a constant contention window. Thus, it does not exponentially increase contention window with every round of backoff. The rationales for keeping the contention window constant are as follows. First, if the RTS failure is due to collision, the contention will likely dissipate because other contending nodes may contend for the channel in other directions as a result of finding unblocked packets after rescanning their queues. Second, if the CTS is not returned due to deafness of the receiver, the binary exponential backoff mechanism usually prolongs the deafness-related delay. Using the same rationales, constant backoff mechanism is also employed in case of retransmission caused by missing ACK. Thus, the idle backoff in OPDMAC is substantially different from the backoff of IEEE 802.11 in two ways: (a) it occurs rarely after finding no unblocked packet instead of transmission failure, and (b) it employs constant backoff time instead of rounds of exponential backoff periods. This is another novel feature of OPDMAC, which is introduced to minimize the impact of backoff period (control overhead) on the throughput and delay of the node.

4.5.2.5 The Listening Period

After each successful transmission, the node is forced to remain idle for a certain period of time called the Listening Period (LP) even if it has packets outstanding for transmission. During the LP, the node listens in an omni-directional mode. The LP is essential to mitigate persistent deafness by allowing other nodes to communicate with the deaf node. Also, overhearing the medium is beneficial because the node needs to collect useful information about its neighborhood to retrieve the channel state information. For example, it has to update its DNAV table which is likely to be outdated as a result of previous beamforming. Although this idle period trades off the spatial reuse, it is necessary to eliminate persistent deafness. In contrast to other directional MAC protocols that employ omni-directional exponential backoff after each transmission failure, the LP, that is derived from a constant window, is needed after each successful transmission. At the end of the LP, the FIFO policy is reinforced. The node scans the packets in the order of their arrival time and transmits the first unblocked packet as mentioned in Sect. 4.5.2.1.

4.5.3   Performance Evaluation

In this section, we evaluate the performance of our OPDMAC protocol. We use OPNET 12.1 [34] as our network simulator. We implemented a smart antenna with directional gain of 10 dB and beamwidth 60° in OPNET using its powerful antenna pattern editor. We also implemented several directional MAC protocols using OPNET Modeler. To focus on the benefits of the spatial reuse, we set the communication range for both directional and omni-directional protocols to 250 m. The packet size is 1024 bytes and the data rate is 11 Mbps. We do not consider node mobility in our simulations.

4.5.3.1   Random Topologies

In the next set of experiments, we evaluate the performance of the OPDMAC protocol in a large multi-hop network. We compare OPDMAC with Basic DMAC protocol, DMAC protocol with omni-directional backoff (DMAC-OM-BO), Circular Directional RTS (CDR) MAC protocol [27] and the IEEE 802.11 standard. In a random network, the challenges are more complex but the additional transmission opportunities can provide more spatial reuse gain. We simulated a network with 30 nodes randomly placed in an area of 1000 m × 1000 m. The results are averaged over 10 different simulation runs. We evaluate the performance for both one-hop flows and multi-hop flows.

In this subsection, we evaluate the performance of the MAC protocols in the presence of multi-hop flows. We consider five CBR flows with random source-destination pairs. The flows are routed over minimum hop routes that are statically assigned. We consider four performance metrics which are the aggregate end-to-end throughput, the average end-to-end delay, the control overhead and Jain's fairness index [21].

$$\text{Fairness Index} = \frac{\left(\sum_{i=1}^{l} x_i\right)^2}{l \sum_{i=1}^{l} x_i^2}, \tag{4.1}$$

where l is the number of flows and x_i is the end-to-end throughput of flow i.

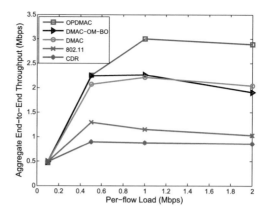

Fig. 4.6 Aggregate
end-to-end throughput for
random multi-hop topologies
with five multi-hop flows

In Fig. 4.6, we plot the aggregate end-to-end throughput versus the per-flow offered load. As we can see, the OPDMAC protocol significantly outperforms the other protocols since it fully exploits the benefits of spatial reuse introduced by the beamforming antennas. We also notice DMAC and DMAC-OM-BO performs better than IEEE 802.11 since they benefit from the spatial reuse although they suffer from deafness while CDR-MAC fails to exploit the benefits of beamforming antennas due to the large overhead used to address their challenges.

Figure 4.7 shows the average end-to-end delay versus the offered load. As expected, CDR-MAC and IEEE 802.11 have the largest delay. DMAC-OM-BO experiences large delay due to its omni-directional backoff that limits the spatial reuse. Our OPDMAC protocol has the smallest end-to-end delay due to its novel backoff mechanism that minimizes the idle waiting time and eliminates the HoL blocking. We also notice that DMAC experiences an average delay that is smaller than DMAC-OM-BO. The reason is that some flows are completely blocked due to persistent deafness. In DMAC, if an intermediate node on the route of a certain flow is also the originator of a new flow, the first flow is blocked as the intermediate node

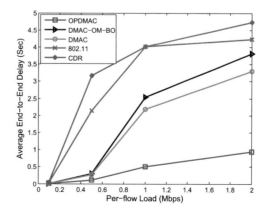

Fig. 4.7 Average end-to-end
delay for random multi-hop
topologies with five
multi-hop flows

Fig. 4.8 The fairness index
for random multi-hop
topologies with five
multi-hop flows

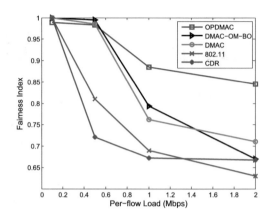

remains deaf as long as its own flow has packets to send. This results in fewer active
flows in the network experiencing a relatively lower delay.

In Fig. 4.8, we plot the fairness index versus the per-flow offered load. As we can
see, the OPDMAC is the fairest among the protocols we compared it with. This is
because OPDMAC protocol reduces the impact of deafness and does not rely on the
binary exponential backoff mechanism, rather it employs a constant window for the
listening period.

Figure 4.9 shows the control overhead. The overhead is defined as the average
number of bits transmitted to deliver one bit of payload to the receiver at the MAC
layer. We can see that CDR-MAC has large overhead due to the circular transmission
of RTS packets. DMAC has slightly more overhead than the rest of protocols since it
suffers from more transmission failures due to deafness. OPDMAC has small control
overhead similar to DMAC-OM-BO and IEEE 802.11. This proves that OPDMAC is
a lightweight protocol that is able to enhance the spatial reuse and reduce the impact
of deafness without additional control overhead.

Fig. 4.9 Control overhead
for random multi-hop
topologies with five
multi-hop flows

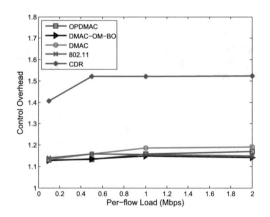

References

1. Arora A, Krunz M, Muqattash A (2004) Directional medium access protocol (DMAP) with power control for wireless ad hoc networks. In: IEEE global telecommunications conference (GLOBECOM), vol 5. Dallas, Texas, pp 2797–2801
2. Bandyopadhyay S, Hasuike K, Horisawa S, Tawara S (2001) An adaptive MAC protocol for wireless ad hoc community network (WACNet) using electronically steerable passive array radiator antenna. In: IEEE global telecommunications conference (GLOBECOM). San Antonio, Texas, pp 2896–2900
3. Bazan O, Jaseemuddin M (2008) An opportunistic directional MAC protocol for multihop wireless networks with switched beam directional antennas. In: IEEE international conference on communications (ICC). Beijing, China, pp 2775–2779
4. Bazan O, Jaseemuddin M (2010) Performance analysis of directional CSMA/CA in the presence of deafness. IET Commun 4(18):2252–2261
5. Capone A, Martignon F, Fratta L (2008) Directional MAC and routing schemes for power controlled wireless mesh networks with adaptive antennas. Elsevier J Ad Hoc Netw 6(6):936–952
6. Chang JJ, Liao W, Hou TC (2009) Reservation-based directional medium access control (RDMAC) protocol for multi-hop wireless networks with directional antennas. In: IEEE international conference on communications (ICC). Dresden, Germany, pp 1–5
7. Chin KW (2007) SpotMAC: a pencil-beam MAC for wireless mesh networks. In: IEEE international conference on computer communications and networks (ICCCN). Honolulu, Hawaii, pp 81–88
8. Choudhury R, Vaidya N (2004) Deafness: a MAC problem in ad hoc networks when using directional antennas. In: IEEE international conference on network protocols (ICNP). Berlin, Germany, pp 283–292
9. Choudhury R, Vaidya N (2007) MAC-layer capture: a problem in wireless mesh networks using beamforming antennas. In: IEEE sensor, mesh and ad hoc communications and networks (SECON). San Diego, California, pp 401–410
10. Choudhury R, Yang X, Ramanathan R, Vaidya N (2002) Using directional antennas for medium access control in ad hoc networks. In: ACM international conference on mobile computing and networking (Mobicom). Atalanta, Georgia, pp 59–70
11. Choudhury RR, Yang X, Ramanathan R, Vaidya NH (2006) On designing MAC protocols for wireless networks using directional antennas. IEEE Trans Mob Comput 5(5):477–491
12. Dai HN, Ng KW, Wu MY (2007) A busy-tone based MAC scheme for wireless ad hoc networks using directional antennas. In: IEEE global telecommunications conference (GLOBECOM). Washington, USA, pp 4969–4973
13. ElBatt T, Anderson T, Ryu B (2003) Performance evaluation of multiple access protocols for ad hoc networks using directional antennas. In: IEEE wireless communications and networking conference (WCNC), vol 2. New Orleans, Louisiana, pp 982–987
14. Fahmy N, Todd T (2004) A selective CSMA protocol with cooperative nulling for ad hoc networks with smart antennas. In: IEEE wireless communications and networking conference (WCNC). Atlanta, Georgia, pp 387–392
15. Fakih K, Diouris JF, Andrieux G (2009) Beamforming in ad hoc networks: MAC design and performance modeling. EURASIP J Wirel Commun Netw (2009), 15 pages, Article ID 839421. https://doi.org/10.1155/2009/839421
16. Gossain H, Cordeiro C, Agrawal DP (2005) MDA: an efficient directional MAC scheme for wireless ad hoc networks. In: IEEE global telecommunications conference (GLOBECOM), vol 6. St. Louis, Missouri, pp 3633–3637
17. Haas ZJ, Deng J (2002) Dual busy tone multiple access (DBTMA) a multiple access control scheme for ad hoc networks. IEEE Trans Commun 50(6):975–985
18. Hsu J, Rubin I (2006) Performance analysis of directional CSMA/CA MAC protocol in mobile adhoc networks. In: IEEE international conference on communications (ICC), vol 8. Istanbul, Turkey, pp 3657–3662

19. Huang Z, Shen C, Srisathapornphat C, Jaikaeo C (2002) A busy-tone based directional MAC protocol for ad hoc networks. In: IEEE military communications conference (Milcom), vol 2. Anaheim, California, pp 1233–1238

20. IEEE (1999) IEEE 802.11 standard: wireless LAN medium access control (MAC) and physical layer (PHY) specification

21. Jain R, Chiu D, Hawe W (1984) A quantitative measure of fairness and discrimination for resource allocation in shared computer system. DEC Technical Report 301

22. Jakllari G, Broustis I, Korakis T, Krishnamurthy SV, Tassiulas L (2005) Handling asymmetry in gain in directional antenna equipped ad hoc networks. In: IEEE international symposium on personal, indoor and mobile radio communications (PIMRC). Berlin, Germany, pp 1284–1288

23. Jakllari G, Luo W, Krishnamurthy SV (2007) An integrated neighbor discovery and MAC protocol for ad hoc networks using directional antennas. IEEE Trans Wirel Commun 6(3):11–21

24. Ko Y, Shankarkumar V, Vaidya N (2000) Medium access control protocols using directional antennas in ad hoc networks. In: IEEE international conference on computer communications (INFOCOM). Tel Aviv, Israel, pp 13–21

25. Kolar V, Tilak S, Abu-Ghazaleh NB (2004) Avoiding head of line blocking in directional antenna. In: IEEE international conference on local computer networks (LCN). Zurich, Switzerland, pp 385–392

26. Korakis T, Jakllari G, Tassiulas L (2003) A MAC protocol for full exploitation of directional antennas in ad-hoc wireless networks. In: ACM international conference on mobile computing and networking (MobiHoc). Annapolis, Maryland, pp 98–107

27. Korakis T, Jakllari G, Tassiulas L (2008) CDR-MAC: a protocol for full exploitation of directional antennas in ad hoc wireless networks. IEEE Trans Mob Comput 7(2):145–155

28. Kulkarni SS, Rosenberg C (2005) DBSMA: a MAC protocol for multi-hop ad-hoc networks with directional antennas. In: IEEE international symposium on personal, indoor and mobile radio communications (PIMRC), vol 12. Berlin, Germany, pp 1371–1377

29. Li Y, Li M, Shu W, Wu MY (2007) FFT-DMAC: a tone based MAC protocol with directional antennas. In: IEEE global telecommunications conference (GLOBECOM). Washington, USA, pp 3661–3665

30. Li Y, Safwat AM (2006) DMAC-DACA: enabling efficient medium access for wireless ad hoc networks with directional antennas. In: IEEE international symposium on wireless pervasive computing (ISWPC). Phuket, Thailand, pp 1–5

31. Takata M, Bandai M, Watanabe T (2006) A receiver-initiated directional MAC protocol for handling deafness in ad hoc networks. In: IEEE international conference on communications (ICC), vol 9. Istanbul, Turkey, pp 4089–4095

32. Mundarath JC, Ramanathan P, Veen BDV (2004) NULLHOC : a MAC protocol for adaptive antenna array based wireless ad hoc networks in multipath environments. In: IEEE global telecommunications conference (GLOBECOM), vol 5. Dallas, Texas, pp 2765–2769

33. Nasipuri A, Ye S, You J, Hiromoto RE (2000) A MAC protocol for mobile ad hoc networks using directional antennas. In: IEEE wireless communications and networking conference (WCNC). Chicago, Illinois, pp 1214–1219

34. OPNET Technologies: OPNET Modeler V. 12.1. http://www.opnet.com/

35. Ramanathan R, Redi J, Santivanez C, Wiggins D, Polit S (2005) Ad hoc networking with directional antennas: a complete system solution. IEEE J Sel Areas Commun 23(3):496–506

36. RamMohan VA, Sethu H, Hosaagrahara MR, Dandekar KR (2007) A new protocol to mitigate the unheard RTS/CTS problem in networks with switched beam antennas. In: IEEE international symposium on wireless pervasive computing (ISWPC). San Juan, Puerto Rico, pp 129–134

37. Shihab E, Cai L, Pan J (2008) A distributed directional-to-directional MAC protocol for asynchronous ad hoc networks. In: IEEE global telecommunications conference (GLOBECOM). New Orleans, Louisiana, pp 1–5

38. Shihab E, Cai L, Pan J (2009) A distributed asynchronous directional-to-directional MAC protocol for wireless ad hoc networks. IEEE Trans Veh Technol 58(9):5124–5134

39. Singh H, Singh S (2003) A MAC protocol based on adaptive beamforming for ad hoc networks. In: IEEE international symposium on personal, indoor and mobile radio communications (PIMRC). Beijing, China, pp 1346–1350

40. Singh H, Singh S (2004) Smart-802.11b MAC protocol for use with smart antennas. In: IEEE international conference on communications (ICC). Paris, France, pp 3684–3688

41. Subramanian AP, Das SR (2007) Addressing deafness and hidden terminal problem in directional antenna based wireless multi-hop networks. In: IEEE international conference on communication systems software and middleware (COMSWARE). Bangalore, India, pp 1–6

42. Takai M, Martin J, Ren A, Bagrodia R (2002) Directional virtual carrier sensing for directional antennas in mobile ad hoc networks. In: ACM international symposium on mobile ad hoc networking and computing (MobiHoc). Lausanne, Switzerland, pp 183–193

43. Takata M, Bandai M, Watanabe T (2007) A MAC protocol with directional antennas for deafness avoidance in ad hoc networks. In: IEEE global telecommunications conference (GLOBECOM). Washington, USA, pp 620–625

44. Takata M, Nagashima K, Watanabe T (2004) A dual access mode MAC protocol for ad hoc networks using smart antennas. In: IEEE international conference on communications (ICC). Paris, France, pp 4182–4186

45. Takatsuka Y, Nagashima K, Takata M, Bandai M, Watanabe T (2006) A directional MAC protocol for practical smart antennas. In: IEEE global telecommunications conference (GLOBECOM). San Francisco, California, pp 1–6

46. Takatsuka Y, Takata M, Bandai M, Watanabe T (2008) A MAC protocol for directional hidden terminal and minor lobe problems. In: IEEE wireless telecommunications symposium (WTS). Pomona, California, pp 210–219

47. Wang J, Fang Y, Wu D (2005) SYN-DMAC: a directional MAC protocol for ad hoc networks with synchronization. In: IEEE military communications conference (Milcom), vol 4. Atlantic City, New Jersey, pp 2258–2263

48. Wang J, Zhai H, Li P, Fang Y, Wu D (2009) Directional medium access control for ad hoc networks. Springer Wirel Netw 15(8):1059–1073

49. Wang Y, Garcia-Luna-Aceves JJ (2003) Collision avoidance in single-channel ad hoc networks using directional antennas. In: IEEE international conference on distributed computing systems (ICDCS). Providence, Rhode Island, pp 640–649

50. Zhang Z (2005) Pure directional transmission and reception algorithms in wireless ad hoc networks with directional antennas. In: IEEE international conference on communications (ICC). Seoul, Korea, pp 3386–3390

51. Zhang Z, Ryu B, Nallamothu G, Huang Z (2005) Performance of all-directional transmission and reception algorithms in wireless ad hoc networks with directional antennas. In: IEEE military communications conference (MILCOM). Atlantic City, New Jersey, pp 225–230

Chapter 5
Enhanced MAC for Millimeter Wave Communication

Abstract Millimeter wave (mmWave) with large unlicensed spectrum is considered the most promising frequency band for next generation multi gigabit per second (Gbps) communication. To address emerging applications and services that demand very high throughput, IEEE 802.11ad and IEEE 802.11ay operating on 60 GHz mmWave band are the two key probable wireless local area network (WLAN) standards. IEEE 802.11ay standard still under the development phase and many of the research groups from academia and industry are working on the task group. IEEE 802.11ad is the amendment of 802.11 to provide data rate of up to 7 Gbps. It operates at frequency band 60 GHz coexisting with 2.4 and 5 GHz and is the newest member of IEEE 802.11 family. In this chapter we first provide a brief overview of mmWave communication focusing on WLAN. We then discuss some novel techniques adopted in IEEE 802.11ad and IEEE 802.11ay standards. Finally, we present three key medium access control (MAC) protocols to overcome the challenges of the of directional mmWave communications.

5.1 Introduction

The millimeter wave (mmWave) spectrum has attained a significant attention to achieve the goal of next generation wireless networks because of its high bandwidth availability. Millimeter wave communication span a wide range of high frequency from 30 to 300 GHz. It opens the opportunity of multi-Gbps communication for consumer wireless local area networks. Therefore, to realize the multi-Gbps communication, IEEE 802.11ad task group was formed to work on 60 GHz band and define modifications on 802.11 physical and MAC layer. IEEE 802.11ad is the latest standard of IEEE 802.11 family that supports applications such as high definition video distribution, instant wireless synchronization etc. To meet the ever-increasing demand of data rate and diverse applications more advance technologies are needed. Hence, IEEE 802.11ay task force was formed to support 20 Gbps and above communication using the unlicensed 60 GHz band, which is developing based on IEEE 802.11ad [3].

© The Author(s), under exclusive license to Springer Nature Switzerland AG 2021   67
O. Bazan et al., *Beamforming Antennas in Wireless Networks*,
SpringerBriefs in Electrical and Computer Engineering,
https://doi.org/10.1007/978-3-030-77459-2_5

Although theoretically mmWave communications can offer multi Gbps data rate with large available spectrum, working on 60 GHz band, several practical challenges are faced due to its unique properties. One of the key characteristics of mmWave channel is high propagation loss compare to systems with 2.4 and 5 GHz bands. The free-space path loss at 60 GHz is about 21–22 dB worse than 5 GHz [7]. Besides the high available bandwidth of 60 GHz, very small wavelength allows implementation of a large number of antenna elements. Efficient use of this spectrum requires a fundamental transition from omnidirectional to direction wireless medium usage. Directional narrow beam antennas can extend the communication range and improve the capacity by means of spatial reuse. Therefore, beamforming using antenna arrays is a way to overcome the increased loss at higher frequencies, which boosts the achievable directivity gain. Some of the key features of IEEE 802.11ad and IEEE 802.11ay are as follows [2, 5, 10]:

- Operates in the millimeter wave frequency of 60 GHz band.
- Both IEEE 802.11ad and IEEE 802.11ay enabled devices support multi gigabit (7–100 Gbps) data rate.
- IEEE 802.11ad and IEEE 802.11ay enabled devices can communicate legacy 802.11 devices, and transparently switch back and forth among 60, 5 and 2.4 GHz, know as multi-band operation.
- May provide Enhanced co-existence among heterogeneous systems such as 802.15.3c WPAN that is working at the 60GHz band.
- Enhanced the 802.11 MAC, by making changes in the beacon interval structure, and supporting both scheduled access and contention-based access.

With directionality, device discovery, spatial reuse, hidden terminal problem and finding the best direction for communication became more difficult than omnidirectional transmission. In Chaps. 3 and 4 we discussed all of the above directional MAC issues in the context of legacy 2.4/5 GHz band. There are two key research directions on the mmWave WLAN. One, is design and update the MAC layer protocols and improve the efficiency of beamforming training (BFT). Second is improving the physical layer such as channel bonding, aggregation, multiple-input and multiple-output (MIMO) etc. Due to the significant attenuation of mmWave as mentioned before, MAC layer design needs considerable changes from lower frequency system. In this chapter we discussed details about the MAC layer schemes for the operation of mmWave WLAN communication focusing on the changes on top of the legacy 802.11. The key challenge of 60 GHz mmWave MAC is handling highly directional transmissions between transmitter and receiver to mitigate the high propagation loss. To support directional multi gigabit (DMG) communication, the IEEE 802.11ad and upcoming IEEE 802.11ay standards introduces a couple of new mechanisms. A novel concept, called "virtual" sectors that discretize the antenna azimuth is adopted. A sector spotlights antenna gains in a particular direction. Therefore, communicating nodes have to agree on the optimal pair of receive and transmit sectors to optimize signal quality and throughput. Another novel concept introduced in mmWave WLAN is the personal basic service set (PBSS). In the PBSS architecture one station take

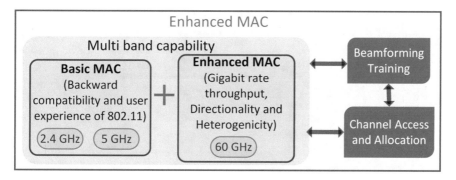

Fig. 5.1 Basic elements of enhanced MAC for millimeter wave communication

the role of PBSS control point (PCP). Only this PCP transmits DMG bacon frames. The key difference between PBSS and independent basic serving set (IBSS) is that, within the IBSS any station can transmit a beacon whereas in PBSS only the PCP is responsible for DMG beacon frame transmission. The PCP allocates the service periods and contention-based access periods and provides the basic timing for the PBSS through DMG beacon. Moreover, PCP also supports QoS and power management. Centralizing these functions at the PCP, directional channel access challenges became more manageable. We will discuss bacon frame format in details later in this chapter [2, 10].

The mmWave MAC layer consist of both basic MAC and enhance MAC to maintain directivity and co-existence as well as to achieve high throughput. Figure 5.1 shows a block diagram of MAC layer presenting the basic component.

Basic MAC provides legacy WLAN functionality and user experience by maintaining the network architecture such as basic service set (BSS), extended service set (ESS), access point and station. The enhance MAC includes new features focusing the mmWave to achieve high throughput. It introduces personal BSS (PBSS) as we mentioned before, support directionality and spatial reuse, and deal coexistence with other 60 GHz standards. The IEEE 802.11ad and 802.11ay also introduced new transmission periods to support multi Gbps throughput. In the next section we will discuss beacon frame format in details for mmWave WLAN.

5.2 Beacon Frame

This section describes the changes to the beacon interval (BI) structure in IEEE 802.11ad and upcoming 802.11ay to deal with the challenges of directional multi gigabit (DMG) communication. The beacon announces the existence of network and carry control information and data. Therefore, channel access is divided into beacon intervals (BI). The BI is subdivided into different phase or access periods and each

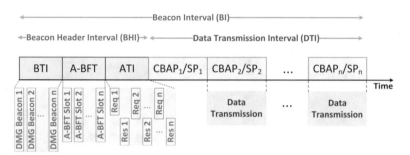

Fig. 5.2 802.11ad beacon interval structure

access period has specific purpose. Figure 5.2 shows the structure of BI in IEEE 802.11ad and 802.11ay [6, 7].

The beacon interval is initially divided into two phases, beacon header interval (BHI) and data transmission interval (DTI). The BHI is used to announce the network and exchange management information to facilitate beamforming training of unasso-ciated DMG and enhanced DMG (EDMG) stations and channel access scheduling. It is comprised of three sub-intervals: beacon transmission interval (BTI), association beamforming training (A-BFT) and announcement transmission interval (ATI).

- **Beacon transmission interval (BTI)**: BTI is the first phase in BHI. In this period an AP/PCP transmit one or more DGM or EDMG beacon frame in different direc-tions or transmit sectors to cover all possible directions. A station willing to join the network scan for a bacon and continue the beamforming process with the PCP/AP.
- **Association beamforming training (A-BFT)**: The second interval is the A-BFT, is used to train the antenna sector between a station and the PCP. This process is the continuity of the DGM beacon transmitted in the beacon transmission interval. The A-BFT period is slotted, and stations randomly select one of the slots. There are up to eight slots is suggested for IEEE 802.11ad and up to 40 slots is suggested for IEEE 802.11ay. During this process, station and PCP launch communication and fine-tune their antenna setting to get better directional data transmission. After the beamforming training is completed the best sector IDs (Sid) are reported for both the transmitter and receiver.
- **Announcement transmission interval (ATI)**: The ATI is used to exchange con-trol and management information between PCP/AP and associated beam-trained stations. This interval uses request and response based transmission allocations for each SPs and CBAPs within the DTI.

The data transmission interval (DTI) is the access period during which stations exchange data frames. The DTI comprises one or more contention-based access periods (CBAPs) and scheduled serviced periods (SPs). The number and order of SPs and CBAPs in the DTI can be any combination.

- **Scheduled serviced periods (SPs)**: Access during SP is assigned for commu-nication between a dedicated pair of nodes as a contention-free period. That is, the ownership of the channel time is granted. Although in the upcoming IEEE 802.11ay, SP is also considered to be allocated for multiple stations using time division duplexing (TDD) that discussed in Sect. 5.4 in details.

- **Contention-based access periods (CBAPs)**: In CBAP multiple stations can participate to access the channel. The IEEE 802.11ad and 802.11ay support both distributed coordination function (DCF) and enhanced distributed channel access function (EDCAF). DCF is the fundamental access method of the IEEE 802.11 MAC that offers fair services for every station in the network. We discussed DCF in Chap. 4. In the following section we discuss about enhanced distributed channel access (EDCA).

5.3 Enhanced Distributed Channel Access (EDCA)

The EDCA provides a contention based differentiated distributed access to the wireless medium. The EDCA defines four access categories (ACs): background, best effort, video and voice. The ACs is used supporting the delivery of traffic with user priority (UPs) at the station. There are eight priority levels as follows: priorities 1 and 2 for background traffic (BK); priorities 0 and 3 for best effort traffic (BE); priorities 4 and 5 for video traffic (VI); and priorities 6 and 7 for voice traffic (VO) [6]. The EDCA allows the high priority ACs to access the channel faster with less backoff waiting time. It uses minimum and maximum contention window size (CW), transmission opportunity (TXOP) and arbitration inter-frame space (AIFS) parameters to define backoff waiting time. A station desiring to initiate transfer of data has to sense the channel idle for a duration of AIFS[AC] instead of same value of DIFS duration in DCF, before starting the backoff countdown. The duration of AIFS[AC] is defined as follows [6]:

$$\text{AIFS[AC]} = \text{SIFS} + \text{AIFSN[AC]} \times \text{slot duration} \qquad (5.1)$$

where, SIFS is the short inter-frame space and AIFSN[AC] is a positive number. The number must be greater than or equal to 2 according to access categories stated above as suggested in the IEEE specification [6]. Slot duration is a basic time unit of duration 5 μs. The contention window CW is used to determine the maximum possible backoff value.

A station generates a random backoff [0, $CW_{AC,min}$], where contention window ($CW_{AC,min}$) size is initialized based on the access categories. As a result, traffic of different priorities to back off for different time intervals. After the AIFS[AC] period, backoff counter starts decrementing by one for every idle slot time, when it reaches to zero, the STA starts the packet transmission using the selected sector in A-BFT interval.

Figure 5.3 shows the simplified backoff procedure of EDCA. Here AC3 represents higher priority and AC0 represents lower priority. Finally, the TXOP period allows a station for the consecutive transmission of multiple frames after achieving channel

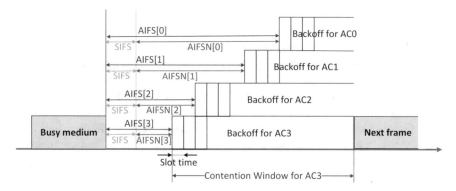

Fig. 5.3 IEEE 802.11ad EDCA channel access method

access. Therefore, a station that wins an EDCA contention, obtain a granted period of time (TXOP) to transmit packets. If the packet suffers a collision, the station increments the contention window size until it reaches to the maximum window size ($CW_{AC,max}$), beyond which the packet is discarded.

5.4   Multiple Channel Access

In order to improve the channel utilization and data rate next generation millimeter wave WLAN supports multi channel operation. To facilitate the coexistence of DMG and enhance DMG (EDMG) stations, network announcement and management frames transmit through the primary channel. Therefore, the BHI elements are on the primary channel. The transmission within the DTI access period can use more than one channel. Multiple channel access can be happened through CBAP/SP or frequency multiplexing [5, 10].

- **Multichannel operation through CBAP/SP**: In order to support the multi channel operation, PCP/AP can allocate bonded and/or aggregated channels for EDMG stations. In this chapter we will not discuss detail operation of channel bonding and aggregation. Figure 5.4a shows the possible channel allocation that support coexistence of DMG and EDMG stations. In this figure, allocation 1 is for legacy devices and allocation 2 is for next generation EDMG devices. As shown in the figure, EDMG PCP/AP can allocate multiple channels overlapping in time or can use only the secondary channels for data transmission and primary channel for control and management. In case of allocation 3, both channel bonding and aggregation is possible because allocated channels are adjacent. Allocation 4 is the example of channel aggregation because of the nonadjacent channel allocation. The allocation can be at CBAP or SP. If the allocation at CBAP, the stations should operate backoff procedure before getting access to the channel. On the other hand,

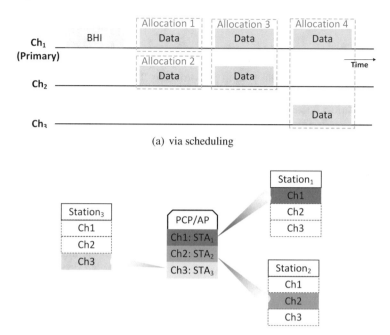

(a) via scheduling

(b) frequency multiplexing

Fig. 5.4 Multiple channel access

if the allocation is SP, DMG stations or EDMG stations execute transmission in the allocated contention free period.

- **Multichannel operation using frequency multiplexing**: To improve the frequency utilization in next generation millimeter wave WLAN the PCP/AP is capable of multi channel operation and allocates each station to different channels using frequency multiplexing. So, multiple stations can access the medium in multiple channels. The intended station communicates through the allocated channel within TXOP period as shown in Fig. 5.4b.

On top of above two multiple access mechanism, next generation millimeter wave WLAN such as IEEE 802.11ay considering a new type of SP named time division duplexing SP (TDDSP). In this approach the SP period is divided into smaller time slots. Each of slots is then assigned to dedicated pair of stations for transmission. The assignment of slots is dynamically done by centralized PCP. Within any assigned slot, a station can only transmit or receive but is not allowed to do both. Therefore, the BI of next generation mmWave WLAN can have both non TDDSP and TDDSP.

Chapter 4 presents the taxonomy of directional MAC protocol as well as explains some key directional MAC protocol in the context of 2.4/5 GHz communication. In the following sections we discuss three major MAC protocols to support DMG and EDMG communication in 60 GHz band.

5.5   Cooperative MAC Protocol

Cooperation at MAC layer gained tremendous attention in the research community and industries to improve the system performance. The cooperative communication enables stations in the same WLAN can help each other to transmit data. The whole idea is, a station, instead of sending frames at a low rate to a destination station directly, takes help from other station or relay station to transmit data at a high rate. The presence of relay station increases the complexity and imposes additional overhead to the network. Moreover, relay selection and resource allocation also became the part of the cooperative MAC design. Cooperative communication technique is adopted in mmWave WLAN to facilitate directional multigigabit transmission [6]. The cooperative MAC protocol for directional multigigabit wireless LAN can be classified into contention based and contention free protocols according to the channel access strategies as we discussed before. The contention free cooperative MAC protocol implements in the scheduled service periods (SPs). Communicating stations do not contend to access the channel, a reserved time slot is allocated for the stations. On the other hand, contention based cooperative MAC protocol implements in the contention-based access periods (CBAPs). Communicating stations contend each other to access the channel for data transmission. The focus of this section is the cooperative MAC protocol for DMG and EDMG transmission.

Figure 5.5 shows a scenario of cooperative communication with relay nodes under a PBSS. There are two types of cooperative communication among the source DMG

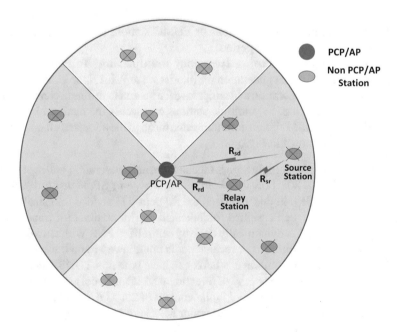

Fig. 5.5 Cooperative communication scenario under the personal basic service set (PBSS)

(SDS) station, destination DMG station (DDS) and relay DMG station (RDS): Link switching and Link cooperation. In link switching, if the direct link between SDS and DDS is disruptive, the SDS redirects the frame to the RDS and RDS forward that to the DDS. In case of link cooperation, an SDS sends frames directly to the DDS and RDS also simultaneously sends frames to the DDS. This simultaneous transmission of same frame will increase the signal quality received at the DDS. Both of the approaches are considered contention free and implement in the SPs [4].

5.5.1 Contention Free Cooperative MAC Protocol

The contention free cooperative MAC protocol implements in the scheduled service periods (SPs), when communicating devices allocated a dedicated time slot. After successful allocation of SP, the allocated SP period is divided into cooperative frame transfer period (CFT) and each CFT is divided into two time intervals T_1 and T_2, where T_1 is the first time interval and T_2 is the second time interval [6]. Figure 5.6 shows the time intervals and the data transmission rules for cooperative communication in the SP.

At the start of each T_1, the source DMG station sends a frame to the RDS. The frame also contains the transmitter address (TA) and receiver address (RA) fields in the MAC header. The TA and RA fields in the MAC header set the MAC address of the SDS and DDS. From the start of the T_2, RDS forwards the frame received frame SDS to the DDS in case of link switching approach. In the link cooperation approach, from the start time of T_2, RDS forward the received frame to the DDS and at the same time (from the start time of T_2) SDS retransmit the same frame to DDS. Therefore, the destination DDS can take the advantage of the improved received signal level from both of the received frame of same original frame. In this case during the T_2, the DDS should set its received antenna pattern such that it simultaneously covers the links between DDS and SDS, and between DDS and RDS. To use the relay node as discussed above, the link quality between SDS and RDS, RDS and DDS, and SDS and DDS need to know. The DMG station learn this information from the link measurement report (LMR) and make sure that link quality of SDS-RDS and

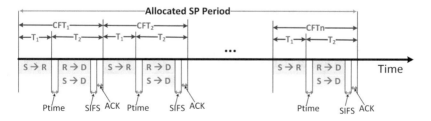

Fig. 5.6 Cooperative data transmission in SP

RDS-DDS is higher than the link quality of direct SDS-DDS, before starting the cooperative communication [6].

5.5.2 Contention-Based Cooperative MAC Protocol

The contention based cooperative MAC protocol implements in the contention based access periods (CBAPs), when stations contend with each other to get access of the channel for transmitting data. The directional cooperative MAC (D-CoopMAC) protocol is a contention based MAC protocol for DMG stations [4, 8]. Similar to the contention free cooperative MAC protocol discussed above, D-CoopMAC also employs relay station to help the source station in data transmission. To find the best relay station, D-CoopMAC protocol introduced cooperation table and relay selection criteria. The cooperation table stores the information of the potential relay stations for every source station. The PCP/AP is involved in the beamforming process between any pair of stations with in the PBSS. Therefore, it is easy to maintain a cooperative table in the PCP/AP. A weight factor is defined to select the best relay station as follows [8]:

$$W_{sr,rd,sd} = \frac{R_{sd}^{-1}}{R_{sr}^{-1} + R_{rd}^{-1}}, \tag{5.2}$$

where, R_{sd} is the data rate of the direct link between source station to PCP/AP. R_{sr} and R_{rd} are the data rates of the two-hop links from the source station to the relay station and relay station to the PCP/AP. A station can include in the candidate relay stations set if $W > 1$. A candidate relay station with highest W value will be selected as a participating relay station.

  If there exist a station that satisfy the above criteria, the cooperation among the source station, relay station and PCP/AC established. Figure 5.7 shows the cooperative transmission process of D-CoopMAC in the directional multi-gigabit communication. The source station sends an RTS frame to the PCP/AP through the beam sector that is selected to communicate with the PCP/AP. The PCP/AP replies with the DMG CTS frame to the source station and the best relay station, indicating whether the cooperative mode is initiated by containing the RA, TA, and HA fields. The HA field contain the MAC address of the best relay station. After that, the relay station sends the DMG Help-To-Send (HTS) frame to both the source station and the PCP/AP. After receiving the DMG HTS frame, the cooperative transmission link is stablished. Subsequently, source station can send its data to relay station and relay station can forward that to PCP/AP [4, 8].

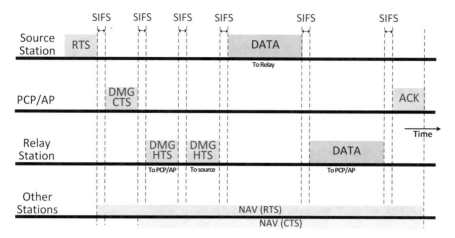

Fig. 5.7 Channel access of directional cooperative MAC in CBAPs

5.6 Dual-Band MAC Protocol

As we mentioned earlier, IEEE 802.11ad and next generation 802.11ay enabled devices can communicate legacy 802.11 devices, and transparently switch back and forth among 60, 5 and 2.4 GHz. To switch among different bands 802.11ad draft specifies fast session transfer (FST) functionality [6]. We are not discussing the FST session transfer process here. This section presents the dual band MAC protocol for DMG communication. The core idea of the dual band MAC protocol is the separation of the control channel and data channel. The control channel will use the omnidirectional lower band 2.4 GHz or 5 GHz and the data channel will use the directional high bandwidth 60 GHz band. Exchanging control message using omnidirectional lower band is more reliable than millimeter band. Every station can sense the control messages and can correctly defer, which may reduce the deafness problem as well. We discussed deafness problem on directional communication in Chap. 4. Moreover, using lower bands for control messages make free resources on the 60 GHz band for DMG and EDMG data transmission.

A station wants to send a data frame, it starts sensing the lower frequency band (2.4 GHz or 5 GHz). If the carrier is idle for a DIFS period, it starts the contention mechanism. The source station sends RTS message to the PCP/AP and PCP/AP reply back the CTS message. Receiving the RTS and CTS message successfully through omnidirectional communication, ensures that the corresponding transmission is know to all stations within the PBSS [9]. The RTS and CTS exchange can occur during the transmission of data frame on 60 GHz band. This can reduce unnecessary delay. The 60 GHz band is exclusively used for data transmission and acknowledgement. Only one station can win a transmit opportunity (TXOP) at a time. Therefore, the next data frame transmission happened only after the previous ACK.

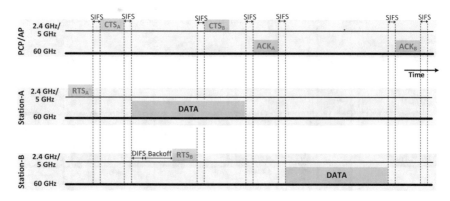

Fig. 5.8 Channel access mechanism of dual-band MAC

The access mechanism of dual-band MAC is shown in Fig. 5.8. In this figure two stations (Station-A and Station-B) wants to transmit data frame using dual band (2.4/5 GHz and 60 GHz) to PCP/AP. The RTS and CTS transmission and backoff is happened in the legacy 2.4 GHz or 5 GHz band and the data frame transmitted in the 60 GHz band [6, 9].

5.7 Pre-defined Virtual Grouping MAC Protocol

This section describes predefined virtual grouping (PDVG) MAC protocol for next generation mmWave WLAN. The PDVG MAC technique utilizes the analog beamforming and spatial multiplexing. The reason to select analog beamforming is to avoid the significant overhead in the digital beamforming and hybrid beamforming. The objective of the PDVP MAC protocol is to mitigate the concurrent link interference and achieve downlink concurrent transmission without adding extra computational overhead [1].

In this MAC protocol, a predefined orthogonal grouping is considered for stations within the PBSS. A number of virtual beam sectors is defined and then stations will be selected from the virtual sectors for concurrent transmission based on best sector ID report of each station. Figure 5.9 shows an example of virtual beam sectors, where coverage area is divided into eight virtual beam sectors ($VS = 8$) with 2 groups. The sector IDs (S_{id}) for each group are selected based on the following equations [1].

$$\left(\frac{VS + 2ij - 2i}{VS}\right) \leq S_{id} \leq \left(\frac{2ij - i}{VS}\right) \tag{5.3}$$

$$\left(\frac{VS + 2ij - i}{VS}\right) \leq S_{id} \leq \left(\frac{2ij}{VS}\right) \tag{5.4}$$

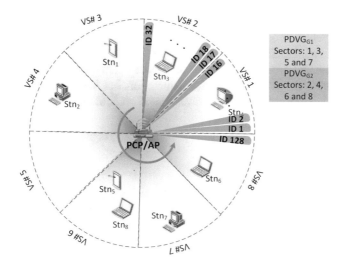

Fig. 5.9 Example of concurrent transmission selection in PDVG

where i is the total number of defined S_{id}, and j is $1 \leq \frac{VS}{2}$.

After the association of each station at the A-BFT, the sector ID is reported to the PCP. For concurrent transmission PCP/AP select one of the groups in coordinated manner. For example, any station that is in the group 1 ($PDVG_{G1}$) is selected for concurrent transmission at the first BI whereas any station in the group 2 is scheduled in the next BI. If more than one station is located in the same selected VS, that is non spatial orthogonal problem exist, scheduling algorithm is needed to select one. Lowest path loss selection can be used to select one station among multiple station in the same virtual sector. The path loss model for a conference room for mmWave as follows [1]:

$$PL[dB] = A + 20 \log_{10}(f) + 10\gamma \log_{10}(d) \tag{5.5}$$

where $A = 45.5$, $f = 60\,\text{GHz}$, pathloss exponent $\gamma = 1.4$ and d is the distance.

Figure 5.10 shows an example of the time allocation for groups in bacon interval. All the stations in $PDVG_{G1}$ is allocated at the first bacon interval and all the stations in $PDVG_{G2}$ is allocated at the second bacon interval. Therefore, every BI has allocated with different group than the previous group. This ensure the concurrent transmission improving the system throughput for next generation WLAN.

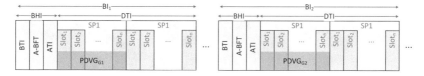

Fig. 5.10 Time allocation for PDVG groups

In this chapter we present wireless local area networks that brings users to the millimeter wave communication. The MAC protocols and issues of two most expected WLAN standard, IEEE802.11ad and IEEE802.11ay for multigigabit communications are presented here. Both of these standards operate on 60 GHz mmWave and support legacy 2.4/5 GHz with multiband operation. In Chap. 8 we will discuss some open research issues related to mmWave WLAN technologies.

References

1. Aldubaikhy K, Shen Q, Wang M, Wu W, Shen X, Osama AM, Yan X, Rob S, Edward A (2017) MAC layer design for concurrent transmissions in millimeter wave WLANs. In: IEEE/CIC international conference on communications in China (ICCC), Qingdao, China
2. Charfi E, Lamia C, Lotfi K (2013) PHY/MAC enhancements and QoS mechanisms for very high throughput WLANs: a survey. IEEE Commun Surv Tutor 15(4):1714–1735
3. Chen C, Oren K, Claudio RCM da S, Carlos C (2019) Millimeter-wave fixed wireless access using IEEE 802.11ay. IEEE Commun Mag 57(12):98–104
4. Chen Q, Jiqiang T, David Tung Chong W, Xiaoming P, Youguang Z (2013) Directional cooperative MAC protocol design and performance analysis for IEEE 802.11 ad WLANs. IEEE Trans Veh Technol 62(6):2667–2677
5. Ghasempour Y, da Silva CR, Cordeiro C, Knightly EW (2017) IEEE 802.11ay: next-generation 60 GHz communication for 100 Gb/s Wi-Fi. IEEE Commun Mag 55(2):186–192
6. IEEE Standards Association (2012) IEEE Std 802.11 ad-2012, Part 11: wireless LAN medium access control (MAC) and physical layer (PHY) specifications, amendment 3: enhancements for very high throughput in the 60 GHz band. In: IEEE Computer Society
7. Nitsche T, Carlos CABF, Edward WK, Eldad P, Joerg CW (2014) IEEE 802.11 ad: directional 60 GHz communication for multi-Gigabit-per-second Wi-Fi. IEEE Commun Mag 52(12):132–141
8. Sami M, Nor KN, Mehdi K, Fazirulhisyam H, Shamala S (2016) A survey and taxonomy on medium access control strategies for cooperative communication in wireless networks: research issues and challenges. IEEE Commun Surv Tutor 18(4):2493–2521
9. Sim GH, Thomas NJCW (2016) Addressing MAC layer inefficiency and deafness of IEEE 802.11ad millimeter wave networks using a multi-band approach. In: IEEE 27th annual international symposium on personal, indoor, and mobile radio communications (PIMRC), Valencia, Spain
10. Zhou P, Kaijun C, Xiao H, Xuming F, Yuguang F, Rong H, Yan L, Yanping L (2018) IEEE 802.11ay-based mmWave WLANs: design challenges and solutions. IEEE Commun Surv Tutor 20(3):1654–1681

Chapter 6
Directional Routing

Abstract In order to fully exploit the benefits of beamforming antennas in multi-hop wireless networks, directional-aware routing protocols should be carefully designed. In this chapter, we discuss several beamforming-related issues and their impact on the routing performance. We also explore different optimization techniques to improve directional routing. Finally, we explain different directional routing schemes that are proposed for multi-hop wireless networks with beamforming antennas.

6.1 Introduction

Routing in a multi-hop wireless network of nodes equipped with Smart Beamforming Antennas is aimed at discovering routes that take advantage of spatial reusability and range extension of smart antennas. However, it also faces the risk of missing the discovery of optimal route due to deafness in directional transmission. Routing in general involves Route Discovery and Route Computation phases. A sub-set of routes are discovered in the route discovery phase that is used in the computation phase to select an optimal route. Routing in a multi-hop wireless network is broadly categorized as Proactive and On-demand (or Reactive).

In proactive routing, routes are discovered through periodic transmission of Route Update messages, similar to IGP routing in wired network (e.g. RIP and OSPF). Every node computes routes to all destinations in the network based on the topology learnt through route update messages. In contrast, routes are discovered as they are needed by their respective source node. In an on-demand routing protocol, e.g. DSR, source node broadcasts route request (RREQ) packet containing the target address (destination) for which the route is sought. When an intermediate node receives RREQ packet for the first time, it appends its node-id in the route and broadcasts the packet. Thus, RREQ packet is flooded in the network. The destination collects the route from the RREQ packet and sends back route reply (RREP) to the source on the reverse route. The DSR is a source routing protocol where every data packet carries the route to the destination. When an intermediate node finds next hop given in the source route unreachable, then it invokes route maintenance procedure. In a

© The Author(s), under exclusive license to Springer Nature Switzerland AG 2021
O. Bazan et al., *Beamforming Antennas in Wireless Networks*,
SpringerBriefs in Electrical and Computer Engineering,
https://doi.org/10.1007/978-3-030-77459-2_6

local repair approach, the intermediate node repairs the downlink by discovering the path to the next hop. In global repair procedure, the intermediate node instead of repairing the failed link sends RERR packet back to the source indicating the failed link. The source after receiving RERR packet may perform a new route discovery. A node receiving RERR packet removes the route corresponding to the failed link.

The exploitation of long range directional transmission in determining shorter routes requires discovery of neighbors at extended range. A node Y is called directional neighbor of node X, if it is reachable only through directional transmission and it is a hidden node for the omni-directional transmission. Otherwise node Y is called omni-directional neighbor since it is also reachable through omni-directional transmission from node X. In a directional routing scheme, directional routes are discovered that take advantage of extended directional transmission range. A directional route is a route that includes at least one link between to directional neighbors. A route that includes no such link is called omni-directional route.

In directional routing, if the set of missing paths contain optimal path, then routing may converge on a sub-optimal path. Routing may miss n number of optimal paths due to a particular deafness condition if all those paths are discovered at the same time. Since proactive routing periodically discovers all paths in the network, it converges to multiple sub-optimal paths. However, all sub-optimal paths may not be in use for carrying traffic. In reactive routing, routes are discovered on-demand basis, hence only fewer sub-optimal paths are discovered but all of them carry traffic.

6.2   Neighbor Discovery

For directional transmission, the transmitter needs to know the receiver as well as the beam to be used for the transmission. For this basic reason a directional MAC protocol keeps track of the neighbors. In a basic neighbor discovery scheme, a node periodically transmits Hello packet, which is when received by another node, then the receiving node discovers the node-id and beam-id in which the Hello packet arrives. Every node maintains a Directional Neighbor Table (DNT), which contains a map of node-id and beam-id. For example, the DNT at node B in Fig. 6.1 shows that it uses beam-1 to communicate with nodes C and D, and beam-4 to communicate with node A. The overhead of Hello packet transmission thus grows linearly with the number of nodes in the network. In another approach [3], a node establishes DNT during route discovery phase of the Directional Routing Protocol (DRP). When the node receives routing control packets, RREQ or RREP, from its neighbor in beam b, then it records beam b and neighbor-id in the DNT as the beam used for the communication with the neighbor. The node can also populate its DNT by overhearing the packets of its neighbors.

In a mobile network, a node may lose track of its neighbor in the beam that is recorded in DNT. This is because the neighbor may move out of its reach in that beam. In a special case where the neighbor moves from the range of one beam to another, the node can rediscover the neighbor through Scanning procedure. In scanning, a node

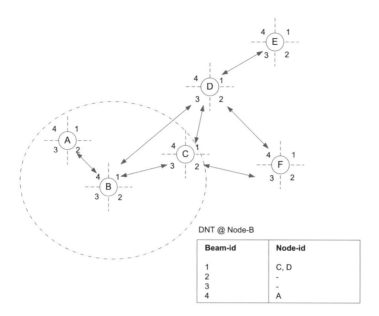

Fig. 6.1 Example of directional routing [3]

transmits Hello packet in every beam soliciting reply from the missing neighbor. When the neighbor receives Hello packet, it immediately replies to the node, which uses reply to determine the new beam that can be used to communicate with the neighbor and updates its DNT. Scanning should be used discretely to avoid its high overhead of generating Hello packets linear to the number of beams b.

6.3 Route Discovery

In a reactive routing protocol, route discovery is performed on demand by flooding RREQ packet in the network. In a simple flooding scheme each node other than the destination broadcasts the packet once. A node can perform Omni-directional Broadcast (OB) through a single communication, but it may reach a shorter distance as compared to transmitting through beamforming. This may cause the source missing shortest-hop route. For example, node A learns route A-B-C-D-E if nodes use omni-directional broadcast for RREQ in Fig. 6.1. Another approach is to perform directional transmission of RREQ in all beams, one beam at a time except in the beam over which the packet is received, which is called Sweeping Broadcast (SB) [1]. It is supported by Directional MAC protocol. In SB, a packet is transmitted in a beam without preceding backoff if the beam is idle, otherwise the beam is marked. In the next round, the packet is transmitted in the marked beams only. If the channel in a marked beam is still found busy, then the beam is skipped without transmitting the packet. This entire procedure is called a single sweep. Skipping a beam may cause missing optimal route, but that is the cost accepted to avoid long delay in sweeping. The SB incurs longer delay than OB because it involves up to b number of transmissions, where b is the number of beams in a node. If SB is used at every node, then node A learns route A-B-D-E that is one hop shorter than A-B-C-D-E in Fig. 6.1. The long delay of SB is amortized over the number of packets that gain from the shorter route. In a routing protocol that employs caching of routes spanning over multiple sessions, the long delay in route discovery attributed to SB will have low impact on average delay.

A Route discovery method is characterized by Control Packet Overhead (CPO), Route Discovery Latency (RDL) and its ability of discovering optimal routes. The SB generates b times more control packets than omni-directional broadcast.

Route Discovery Latency is the time elapsed at the source between sending the RREQ packet and receiving the first corresponding RREP packet. When SB is used to flood the RREQ packet, several factors affect the RDL. The benefit of long range transmission due to directional transmission of RREQ and spatial reuse tend to reduce RDL. In SB while RREQ undergoes transmission in one beam, communications in other beams among other node pairs can occur simultaneously. The impact of long range transmission of RREQ is more pronounced in comparison with OB when the distance of separation of source and destination is large. In small distance separation, both SB and OB tend to discover paths of the same distance (shortest path). It is for the long distance of separation the OB tends to lose to SB in discovering the shortest paths. Conversely, sweeping delay, deafness and interference may increase RDL. A node along the optimal path may not receive RREQ due to deafness or interference, causing the RREQ to traverse over long distance with higher RDL. Interference due to side lobes increases the probability of collision in the main lobe transmission that in turn adds retransmission delay to the RDL. When source and destination nodes are separated by a large distance, the long transmission range of RREQ becomes a dominant factor shadowing the effect of sweeping delay [1]. However, node density may also affect RDL as higher node density leads to higher side-lobe interference causing collisions.

The main reason of a route discovery converging to a sub-optimal path is when only a sub-set of paths are discovered leading to a situation where the optimal path remains in the missing sub-set. The SB tends to discover shorter path as compared to OB due to long range directional transmission of RREQ. However, a node along an optimal path may not receive RREQ due to deafness, which mitigates the gain of long range transmission and results in discovering a sub-optimal path. Another reason of converging to a sub-optimal path in SB is that since an intermediate node forwards only one RREQ packet and drops the subsequent RREQ packets; it may forward the RREQ packet traversing a sub-optimal path.

6.3.1 Route Discovery Optimization

The sweeping broadcast is used to discover shorter-hop optimal routes by considering extended range of directional transmission. There are two sources of inefficiency of SB. One, it generates redundant control packets as one RREQ packet is transmitted in each direction and contributes positively to broadcast storm problem. Two, it adds delay to the route discovery since sweeping may take two rounds to complete visiting every beam as it might revisit some beams when found blocked in the first round. A beam is blocked by DNAV due to an ongoing transmission between neighbors in the direction of that beam. There are two approaches of dealing with SB inefficiency. First, new methods are developed to reduce sweeping redundancy (i.e. CPO) and delay (i.e. RDL). Second approach is to abandon SB and modify OB to compensate its deficiency of shorter transmission range.

Sweeping Broadcast Optimization

The SB can be optimized for reducing redundancy and sweeping delay by restricting the selection of beams in which the RREQ packet is forwarded to only those beams through which the node has not received a RREQ packet [4]. This is called Simple Enhanced Directional Flooding (SEDF). When a node receives a RREQ packet, it triggers a delay timer. If the node receives the same RREQ packet again before the timer expires, keeps a list of the beams over which it receives the same RREQ packet. At the expiration of the delay timer, the node forwards the RREQ packet in only those beams that are not in the list.

In Single Relay Broadcast (SRB) scheme the node forwarding RREQ packet instead of broadcasting the packet in a beam, it selects a single relay node in the beam and unicast the packet to the relay node. It needs the support of neighbor discovery mechanism for this selection. A node periodically sends hello packets in every beam to participate in the neighbor discovery process. When it receives hello packet from its neighbor it measures the receive power (P_r) of that packet. It selects a neighbor as relay in a beam if the P_r of the hello packet it receives from the neighbor is the lowest of the receive power of hello packets from all other neighbors in the same beam. Thus, the node selects among its neighbors in a beam the farthest node as relay, which ensures that the RREQ discovers shortest-hop route. The SRB reduces redundant transmission of RREQ packets in the network and effectively controls the flooding storm. However, it may not be effective in discovering optimal routes, as it limits the discovery of routes only through relay nodes in a beam, which may lead to partitioning the network for route discovery. For example, node A selects node D as relay in beam-1 causing nodes C and F not receiving the RREQ packet. This leads to partitioning of network into A-D-E and C-F in terms of flooding of RREQ packet in the network.

The broadcast storm of SB can be effectively reduced by choosing a single relay in every beam where RREQ is retransmitted. The amount of redundancy can further be reduced by ordering the selection of beams for RREQ retransmission. The New Enhanced Directional Flooding (NEDF) employs both restricted selection of beams for retransmit, as in SEDF, and selection of a single relay in a beam for forwarding RREQ to the next hop, as in SRB. It also introduces ordering in the beam selection for forwarding RREQ. When a node receives RREQ packet for forwarding, it starts the delay timer if the timer is off, and marks the beam passive over which it has received the RREQ packet. It also marks a beam passive if the beam is busy or there is no neighbor in the beam. A beam is considered busy if the DNAV is set to a positive value reflecting ongoing transmission in the beam, thus it is marked as passive. After the timer expires, the node selects the beams from the set of beams that are not marked passive. It begins sweeping from a beam in the selected set that is diagonally opposite to the beam over which RREQ is received. The idea is to select the beams in the order of the beams showing plausible maximum coverage. The diagonally opposite beam usually contains the farthest relay. Then, the beams adjacent to the diagonally opposite beam are selected for transmission of RREQ. In the case when RREQ is received over multiple beams, the node chooses a reference beam in determining the diagonally opposite beam to begin sweeping. It chooses the beam where RREQ packet is received with minimum receive power P_r as the reference beam. The NEDF improves SRB by extending the coverage of RREQ thus further reduce the redundancy. However, it faces with the risk of network partitioning for RREQ reachability in the network.

The inherent problem of SRB and NEDF is the risk of network partitioning that is primarily due to limiting RREQ forwarding to a single relay in a beam. The Probabilistic Relay Broadcasting (PRB) is a scheme that overcomes the inherent problem of SRB (and NEDF) by allowing multiple relays in a beam. The relay selection is distributed, that is instead of identifying relay node in the RREQ packet by the forwarding node, it includes the farthest power, P_f, in the packet, which is used by a node to calculate the probability of forwarding the RREQ packet. When a node receives the same RREQ packet from multiple neighbors in a beam, it records the lowest receive power as the farthest power P_f for RREQ in that beam. The P_f gives a sense of farthest distance between the receiving node and its neighbor from where it receives the RREQ packet. Thus, the node keeps a list of (beam-id, P_f) pairs corresponding to all the beams over which it has received the RREQ packet. After the delay timer expires, the node forwards RREQ packet in all the beams. It includes corresponding P_f values in the RREQ packet transmitted in a beam. When a neighbor receives RREQ packet it calculates its own receive power P_r of the packet. It decides about forwarding the received RREQ packet with a probability, $P_h = P_f/P_r$, where P_h denotes the forwarding probability. The P_h divides the beam into zones of higher and lower probability of relay selection. Hence, relay selection can be further limited to a zone by constraining the value of P_h that is used for forwarding the packet. For example, a node is constrained to forward the RREQ packet only if P_h is in (0.5, 1), that is the relay selection is restricted to the zone in which $P_r = 2 * P_f$ to $P_r = P_f$. For example, Fig. 6.2 shows that nodes 3, 4, and 5 calculate P_h in the given range and become relay in forwarding the packet. Node 4 is the farthest node and has the highest P_h value (1). The drawback of PRB is that with probability $\pi(1 - \max P_h^i)$, though small, there will be no relay selected in a beam.

Fig. 6.2 Example of
PRB [4]

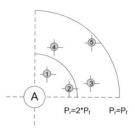

The above optimizations are proposed to primarily reduce the redundancy and sweeping delay of SB. Despite the design objective of SB to find shorter hop routes by taking advantage of extended range of directional transmission, it is prone to miss optimal routes, especially in high load situation [5]. There are several reasons for missing optimal routes. First, deafness and collision can lose transmission in a beam. Second, due to skipping beam, the RREQ packet may never be transmitted in a beam that is blocked due to, for instance, ongoing transmission. Third, sweeping inherently serialize transmission of RREQ and imposes an order on selection of beams for transmission. The neighbors of a node receive packets in that order, which biases the route discovery in favour of those neighbors. For instance, if these neighbors reach an intermediate neighbor x that is on the optimal path before other nodes reach x, then the route discovery fails in converging to optimal path. This is because a node transmits RREQ packet only first time it receives the packet, and in this case the node x receives RREQ through another node. In Fig. 6.1, if node C uses SB of RREQ packet for discovering a route to destination E and the sweeping order it has is 2-1-4-3, then node F receives the packet first. It may succeed forwarding the RREQ to node D before node C is able to forward to node D. The node D in this case sets the backward path for RRES via node F; hence the route discovery at node C converges to a sub-optimal path C-F-D-E. Fourth, nearby neighbors of a node may receive RREQ packet on its side lobes. Since, side lobe transmission cannot reach directional neighbors; the routes discovered through these nearby neighbors in the side lobes may be longer than optimal routes. Further, the forwarding of RREQ packet through these nearby neighbors in the side lobes may block discovery of optimal routes.

Route Compaction

A radical approach is to abandon sweeping broadcast in favour of omni-directional broadcast. And then compact the routes discovered through OB using extended range of directional transmission. The compaction results in shortening the route. The use of OB eliminates all four problems associated with SB, as discussed before. The downside of route compaction is that it fails to discover routes in a sparse network where some nodes are only reachable through direct transmission, since RREQ broadcast cannot reach those nodes through omni-directional flooding. In this section route compaction is discussed in the context of source routing, e.g. DSR.

The route compaction scheme goes through two distinct phases: route discovery and route compaction. During route discovery a node performs omni-directional broadcast of RREQ packet. The result of route discovery phase is a set of routes between a (source, destination) pair consisting of only omni-directional neighbors. In the subsequent route compaction phase, an intermediate node can compact one or more hops into a single hop by replacing one or a chain of omni-directional neighbors into a single directional neighbor. For example, the route A-B-C-D-E can be discovered through OB and then compacted into A-B-D-E in Fig. 6.1. The broadcast of RREQ by node B cannot reach D, because D is only directly reachable through directional transmission from B, hence route discovery only find the path B-C-D from B to D. The compaction of B-C-D into B-D requires directional neighbor discovery. Therefore, route compaction design space has two aspects: (i) Alternative approaches to the implementation of route compaction, and (ii) Directional Neighbor Discovery.

Active Route Compaction

Route compaction can be performed either at an intermediate node along an active path, or at the source. A path is an active path if packet flow between the source and the destination of the path remains in progress. When an intermediate node along the path finds that a downstream node indicated in the source route of the packet two or more hops away from the node is a directional neighbor, and then it replaces the source route in its cache with the compacted route. The compaction of route at an intermediate node is limited to the compaction of only active route. Alternatively, the intermediate node instead of performing compaction of the route in its cache sends Gratuitous Route Reply to the source. This allows the source to either compact the given route or take another shorter route altogether. Thus, the source will always be able to find a shorter and more optimal route, which can potentially be other than the active route; hence it overcomes the limitation of compaction at an intermediate node.

The key component needed for route compaction is the discovery of directional neighbors. The MAC layer support is required in Directional Neighbor Discovery. Since, AoA cache in MAC layer is populated with only overheard neighbors, it is less accurate and incomplete. In Passive Discovery and Compaction (PaDC) scheme AoA cache entries are used in route compaction, which tends to miss compaction opportunities. In contrast, Proactive Discovery and Compaction (ProDC) scheme employs an explicit Neighbor Discovery to populate the AoA cache, which improves accuracy of directional information of neighbor and widens the coverage of neighbors to including more than overheard neighbors in the cache. The node periodically sweep neighbor discovery packet (typically a hello packet) in all beams, thus a receiving node gets information about the sending node that is used to detect directional neighbors for route compaction. The ability of ProDC to achieve effective compaction comes with a higher sweeping cost than SB. The cost of periodic sweeping of ND packets in every beam can be mitigated by limiting the sweeping to clusters of active beams, called Active Sector Optimization for ProDC. The central idea of this approach is the path brings traffic closer to the destination and candidate nodes for compaction tends to cluster in the direction of transmission. Hence, a node measures incoming traffic periodically in all of its beams. When it finds incoming traffic in a beam above than pre-configured threshold, it marks that beam as active beam. It then sends ND packets in active beam and two of the adjacent beams, which are used by the receiving nodes to discover directional neighbor relationship with the node for compaction. The discovery of directional neighbors in active and its adjacent beams is called Active Sector Neighbor Discovery (ASD). For example, after sometime of traffic flow along the path A-B-C-D-E in Fig. 6.1, node D marks beam 3 as active beam. It then send ND packet in beam 3 and the two adjacent beams, 2 and 4. Both nodes B and C receive ND packet in beam 3, which helps node B in discovering the directional hop to D and compacting the route A-B-C-D-E to A-B-D-E. The optimization reduces ND packet overhead by not sending the packet in beam 1. A further optimization is performed in Reactive Discovery Compaction (ReDC) by limiting the compaction to be performed by nodes that form the route in addition to restricting to ND to active beams and beams adjacent to them. The node do compaction on a route can do that for several routes through a single ND if it is on multiple routes. The additional constraint in ReDC reduces significantly the number of ND transmissions.

Any Route Compaction

The route compaction opportunity can be severely constrained if the compaction is only limited along the active path. The optimal route discovered through omni-directional flooding may not necessarily be a good candidate to achieve high degree of compaction. For example, consider grid topology in Fig. 6.3, where each node is separated by a fixed distance with its horizontal and vertical neighbors and a node is reachable through omni-directional transmission from its horizontal, vertical and diagonal neighbors. Further assume two nodes separated by two hops along a straight path are directional neighbors, e.g. node 1 can directly reach node 3 through directional transmission but not through omni-directional transmission. The equal-cost multipath routes from node 7 to node 12 include: (7-8-9-10-11-12), (7-14-21-16-11-12), and (7-2-9-16-17-12) routes. The only route that can be compacted into an optimal directional route (of 3 hops length) is the straight route, (7-8-9-10-11-12). If one of the other two routes is active, then optimal route through compaction cannot be achieved.

The near optimal directional route after compaction can be determined if the input to compaction algorithm includes both optimal and non-optimal omni-directional routes. Since the source can collect all such routes, it is best suited to run the compaction algorithm. A critical component in the implementation of this scheme is the ability of the source to detect those omni-directional routes that can be compacted into optimal or near optimal directional routes. The omni-directional flooding does not discover complete topology [6] since a node participating in flooding rebroadcasts only first broadcast packet and drops the subsequent packets even though they traverse different paths. Hence, an effective route discovery method is another component of the implementation of compaction algorithm. In Multi-Route Attachment (MRA) scheme an intermediate node forwards the first RREQ packet, but keeps a

Fig. 6.3 Example of any
route compaction [5]

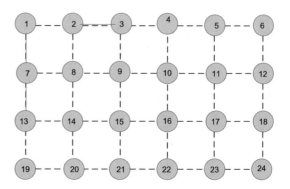

record of all the routes reported in all RREQ packets it receives including those that it did not forward and dropped. When the intermediate node receives RREP from the destination, it includes in the same RREP packet all the routes from its record and forwards RREP packet along the corresponding reverse path. For example, as shown in Fig. 6.3, node 9 receives three RREQ packets from nodes 2, 8 and 14 in that order corresponding to route discovery for source = 7 and destination = 12. It forwards the RREQ it has received from node 2 while record routes of all other RREQ packets as, [s = 7, d = 12, routes = (7-2), (7-8), (7-14)]. When it receives RREP from the destination with the route (7, 2, 9, 10, 11, 12), it attaches routes formed using other two sub-paths and consequently sends three routes on the reverse path to the source in the single RREP path. The three routes it sends are: (7, 2, 9, 10, 11, 12), (7, 8, 9, 10, 11, 12) and (7, 14, 9, 10, 11, 12). It then forwards the RREP packet to node 2. The number of routes an intermediate node manages to include in the RREP packet depends upon how soon it gets RREP after receiving first RREQ. For example, in the above situation node 2 was able to include only three routes in the RREP packet, since RREQs from other nodes, such as nodes 3 and 15 arrived after the arrival of RREP from node 12. This phenomenon tends to occur at intermediate nodes closer to the destination. The MRA scheme does not guarantee discovering all routes, as opposed to the simple approach of forwarding all RREQ packets, but at a much lower cost of forwarding a single RREQ and RREP packet, it discovers a significant percentage of routes that likely includes the route to be compacted to a near optimal directional route.

6.4 Directional Routing Schemes

6.4.1 Directional DSR (DDSR)

In a simple implementation of DSR on DiMAC [1], called Directional DSR (DDSR), RREQ packet is flooded through directional sweeping broadcast. A node maintains DNT to determine the beam-id for communication with its neighbor. The DNT is updated by overhearing packet transmission in every beam. The sweeping allows discovery of optimal directional routes including directional neighbors. In case of missing link with a neighbor, which is identified by retransmit timeout in the given beam, the node employs scanning procedure to rediscover its link with the neighbor. The large separation between nodes accentuate the benefit of shorter hops in directional routes due to extended directional transmission, which tends to offset the drawback of sweeping delay in DDSR. Simulation results show that throughput of DDSR is comparable with DSR and in some situations even lower than that, especially when 4-beam antennas are used. The source of reduced throughput is the convergence of DDSR to sub-optimal routes. Since, destination sends RREP in response to first RREQ it receives, this hasty response includes best route based on a smaller set of routes discovered. The subsequent RREQ packets that are dropped would likely contain optimal directional route.

The DDSR incorporates Delayed Route Reply Optimization [1], which requires the destination node to delay sending RREP after receiving first RREQ. The delay parameter T is calculated as $T = p * T_{sweep}$, where T_{sweep} is the sweeping delay that is the time taken to complete one sweep, and p is a configuration parameter. This delay T allows the destination to choose the best route among all the routes received in RREQ packets within T. The delayed route reply optimization is not equivalent to replying all RREQs. While a node beamforms towards its neighbor to send a RREP packet, it misses receiving all other packets including subsequent RREQ packets due to deafness. Hence, delay T is crucial because it allows the destination collecting routes from several RREQs. The drawback of this optimization is that it adds T to RDL.

6.4.2 Directional Routing Protocol (DRP)

The Directional Routing Protocol (DRP) modifies DSR for directional routing [3]. In DRP, source and all intermediate nodes perform sweeping broadcast of the RREQ packet. The DRP employs NEDF optimization of sweeping broadcast without SRB capability. A node broadcasts in a beam instead of unicasting to a single relay in the beam. The selection of beam in sweeping phase is also ordered and sweeping starts from the diagonally opposite beam. The RREQ contains Directional Route Record (DRR), which is a slightly modified version of the source route record of DSR. The RREQ forwarding node adds beam-id of the beam over which it has received the RREQ packet in addition to its node-id to the directional route record. For example in Fig. 6.1, the RREQ that is received by node E in its beam 3 contains DRR = (A, –), (B, 4), (D, 3). The destination sends RREP along the reverse route formed from DRR.

The DRP includes route maintenance procedure in case of a link failure. The mobility of next hop causes link failure on a route. A node detects link failure by not receiving Acknowledgment to a repeated transmission exceeding retransmit count. The link failure may be due to the mobility of next hop node from one beam to another within the directional transmission range, which can be repaired using a local link repair procedure. The link failure that is caused by the mobility of next hop to the outside transmission range can be recovered through a global recovery procedure. The DRP employs Location Tracking and Two-hop Directional Local Recovery in the local recovery procedure. It also defines Route Recovery for global repair.

Local Route Repair

The local route repair in DRP is achieved in two steps; first through location tracking and then through two-hop directional local recovery. The DRP adopts a two-phase location tracking scheme, where a node after detecting link failure scans the next hop node in adjacent beams. The rationale of scanning in adjacent beams is depicted in Fig. 6.4a. Node Y moving from its original position (x, y) to (x1, y1) in one of the adjacent beams remains along the path to the destination Z. However, if it moves to new position (x2, y2) in a distant beam, then it is no longer along the route to destination Z. The local repair will likely succeed if the new location of the next hop is not too far from the original location, such as in one of the adjacent beams. The number of adjacent beams to be scanned for the next hop depends upon the beam width among other factors. Figure 6.4b shows the angular region that needs to be searched for the new location and the table of adjacent beams that need to be searched for the corresponding beam width. Assuming d be the distance between nodes X and Z and r be the communication radius, then X must scan for Y in the angular region (shaded portion in the Figure) defined by $2 * a * \cos(d/2r)$. For $d = 1000$ and beam width 45°, $\theta = 67°$ that is node X needs to scan in $(i - 1)$ and $(i + 1)$ adjacent beams in addition to the original location of Y in beam i. If beam width changes to 30°, then θ becomes 90° and node X needs to scan Y in $(i - 2)$, $(i - 1)$, $(i + 1)$, and $(i + 2)$ beams in addition to beam i. The location tracking is triggered after RTS transmission fails up to certain number of retransmit attempts. For example, if RTS from X to Y fails in beam i for three consecutive attempts, then node X transmits RTS for the remaining tries (4th, 5th, 6th and 7th) in n adjacent beams, where n is determined as described above. After seven retries of RTS, which is the maximum value of retransmit count configured for RTS, the RTS transmission attempt is aborted and location tracking is assumed to have failed in locating node Y. Then, the two-phase local recovery procedure begins to find a bypass route (up to two hops length) to the second next hop node, that is node Z in case of Fig. 6.4a.

In two-phase local recovery, node X sends directional RREQ to find the route to the second next hop, node Z, such that the scope of RREQ (e.g. TTL = 2) is set to two hops. The intermediate node X determines the second next hop id (=Z) from the DRR in the data packet under transmission. The intermediate node waits up to time T_l for RREP from the second next hop. If it receives RREP within T_l, it sends RERR with $LOC_R ERY = TRUE$ to the source node for the source to update corresponding route. If it fails to receive RREP within T_l, then it sends RERR with $LOC_R ERY$ flag $= FALSE$, which indicates the source to start global route repair.

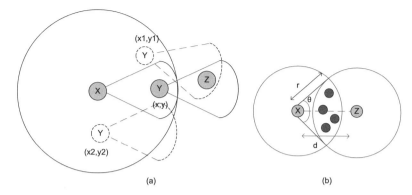

Fig. 6.4 (**a**) Mobility of the node in adjacent zones, and (**b**) estimation of angular region [3]

Global Route Repair

The DRP defines Route Recovery procedure for global route repair. The recovery procedure Zonal Repair that limits the rediscovery of a new route to a zone in the network that is more plausible to contain both source and destination node. The zone is calculated based on the relative directions of nodes from the source along the broken route. For example, in Fig. 6.1, if the separation between nodes A and B is assumed to be half the transmission range, then node A can estimate the coordinates of node B assuming it lies on the angular bisector of beam-2 to be $(R * \cos(\beta/2), R * \sin(\beta/2))$, where R is half the transmission range, and β is the beam width. Similarly, the coordinates of D and E and be calculated relative to their previous nodes, and then estimated relative to A. The angle for the angular zone in which nodes B, D, and E lie is calculated at node A, and it is then padded with $\beta/2$ on each side. The source broadcasts RREQ only in the beams within the calculated zone that is, in this case, within beams 1 and 2 of node A. The zonal repair is efficient in reducing both the redundant RREQ packets and in reducing the RDL. The zonal repair, however, does not guarantee that search for new route always succeed even though the destination is reachable through a connected path. In case of failure of zonal repair in discovering a new route, the source node launches a new route discovery by flooding RREQ packet in the whole network.

6.4.3 Zone Disjoint Routing (ZDR)

Interference in wireless communication due to an ongoing transmission within the transmission range of a node blocks it from communicating with another node. The blocking of a link on the path causes delay in traffic flow. The spatial reusability of directional antenna enhances the opportunity of finding paths with no blocked links.

Fig. 6.5 Zonal disjoint
routing [7]

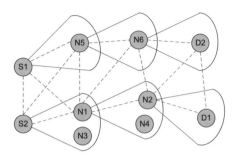

Zone Disjoint Routing (ZDR) exploits the spatial reusability offered by directional
antennas in route computation [7].

The delay performance of a route not only depends on the status of congestion
at intermediate nodes but also upon the number of active nodes engaged in ongoing
transmission in the transmission range of a node along the path. Figure 6.5 shows
topology of a network that connects source nodes S1 and S2 with destination nodes
D1 and D2 respectively. Two paths that have no nodes in common are called Node
Disjoint Paths with respect to each other. Node disjoint paths allow load balancing;
hence reduce congestion in the network. For example, Paths, P1 = S1, N1, N2,
D1 and P2 = S2, N3, N4, D2, are node disjoint paths, hence provide load balancing
across the paths. Consider a flow from S1 to D1 is already established on the path P1.
Since ongoing transmission on link S1–N1 blocks the transmission on link S2–N3
assuming omni-directional antennas, path P2 despite of being a node-disjoint path
with P1 experiences delay due to blocking of links. The other node-disjoint path, P3
= S2, N5, N6, D2 experiences the same delay due to blocking of its links caused
by ongoing transmission on P1. This blocking of links on a path p due to ongoing
transmission on another path q is called Route Coupling. A path carrying packets of
a flow is called Active Path. In directional antenna all the nodes within a beam are
blocked if there is a transmission in that beam. But, nodes outside the beam remain
unblocked. If we consider directional transmission in the network, then nodes N3 and
N4 are blocked by because they are in the S1–N1 and N1–N2 beams respectively.
Consequently, paths P1 and P2 induce route coupling related delays in their packet
flows. However, paths P1 and P3 show no route coupling due to spatial reusability
of directional antenna. Analysis of P3 reveals that P3 is not only node-disjoint path
but also its nodes are not in the beam of any link of P1. Two paths such that none of
their nodes are in the transmission beam of any link of path P2, then the two paths
are said to be Zone Disjoint.

The aim of ZDR is to discover zone disjoint path for a given flow f between
source-destination pair (Si, Di) while a set of flows F exists in the network. In ZDR,
every node in the network keeps track of the nodes undergoing active transmission
among its neighbors, as well as an estimate of node activity across the whole network.
Route computation at every node in ZDR evaluates route coupling of a path with
respect to active paths in the network based on the estimate of its node activity in
the network. In ZDR, a beam α at a node n of width β and range R is called a

transmission zone (α). The correlation factor η_n of a node n along the path P is the set of active directional neighbors in the zone (α), which is due to its beam in the direction of the next hop node m along the path P. The correlation factor $\eta(P)$ of path P is the sum of correlation factors of all the nodes on the path. When $\eta(P) = 0$ of a path P, then it is knows as zone disjoint path with respect to all active paths in the network. Hence, $\eta(P)$ is used as a path metric in ZDR for evaluating a path in route computation.

The ZDR is a table-driven (proactive) routing protocol. Every node in the network maintains three lists: Neighbourhood Active Node List (NANL), Active Node List (ANL) and Global Link-State Table (GLST). It also maintains Directional Neighbor Table (DNT), as discussed in Sect. 6.2, modified to keep the range of receiving beam at node n when it receives transmission from its neighbor m. Each node broadcasts its ANL after periodic interval T_A. It serves two purposes: (i) node n after receiving ANL from all its neighbors i, j and k records their communication status along with the transmission range in its DNT, and (ii) the node also forms NANL, and updates its ANL. Each node also broadcasts its GLST after periodical interval T_B. A node updates its GLST after receiving GLST from its neighbor. In route computation phase, the GLST at node n is used to determine the shortest-path from source S to destination D, and NANL and ANL are used to compute $\eta(P)$. For example, at S2: NANL $=$ S1, N1, ANL $=$ S1, N1, N2, D1, and GLST $=$ P2, P3. It computes $\eta(P2) = 1 + 1 + 0 = 2$, and $\eta(P3) = 0 + 0 + 0 = 0$. Hence, it selects P3 for packet flow to destination D2.

6.4.4 DSR with Directional Antenna for Partition Bridging

One of the benefits of directional antenna is to increase the transmission range while using transmit power of equivalent omni-directional transmission. This capability of directional antenna can be used selectively to bridge network partitions. Node mobility in an ad hoc wireless network can cause network partitioning, which may be permanent or transient. The network partition can be bridged by extending the hop range through increased transmit power. In omni-directional antenna system, increasing transmit power has an adverse effect on increasing interference in the vicinity of the transmitter, which in turn blocks active routes due to route coupling. In directional antenna system, increasing transmit power in a beam extends the range of that beam, but does not impact transmissions in other beams. Hence, directional transmission can be effectively used to bridge network partitions by both with or without increasing transmit power of equivalent omni-directional transmission.

The DSR routing protocol can be modified to bridge network partitions selectively for those packets that experience routing outage [8]. In the normal operation, RREQ packets are broadcast through omni-directional antenna. When a node is unable to locate omni-directional next hop neighbor, it initiates directional transmission of RREQ packet to locate a next hop node in a specific angular region for long hop. The distance of a hop extended beyond omni-directional transmission range is called a

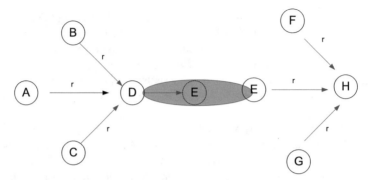

Fig. 6.6 Example of the use of directional antenna in bridging a network partition, node E moves from it original position to the new position and r is the omni-directional transmission range

long hop; otherwise it is called a normal hop. It is assumed that all the nodes within 120° of the angle of arrival of a packet (60° on either side) in a normal hop receive the packet, hence the angular region for long hop is the region bounded by 240° in the opposite direction of the arrival of a packet. The search for long hop neighbor is performed within this angular region (Fig. 6.6).

Route Discovery

In modified DSR, every node maintains a Passive Acknowledgment Table (PAT), which contains following record corresponding to each target address: <target address, record insertion time, list of RREQs, list of angular region>. The source route header of a packet is also modified to include trigger partition bridging flag and long hop flag. The trigger partition bridging flag is set in the RREQ packet that is used by an intermediate node in the network to bridge network partitions, and if the long hop flag is set in a packet, it indicates that the packet is transmitted over long hop with high transmit power. When a source node has a data packet to send and it does not find a route to the destination in its route cache, then it broadcasts using omni-directional transmission the RREQ packet with trigger partition bridging flag reset. If it finds that a RREQ packet was earlier sent for the same destination but the packet has timed out without reply, then it broadcasts a new RREQ packet using omni-directional transmission with trigger partition bridging flag set.

   When an intermediate node receives a RREQ packet and it does not find an entry in the PAT corresponding to the target address, then it creates a new entry and enters in the PAT before forwarding the RREQ packet using omni-directional broadcast. It initializes the angular region in the new entry with all the directions within 240° of the direction opposite to the direction of arrival of the RREQ packet. In the case when the node finds an entry in the PAT corresponding to the target address, it adds the RREQ to the list of RREQs in the entry. It calculates Δ = present time − record insertion time; if the Δ is less than the threshold, then it updates an angular region in the entry by leaving out the overlapping region between the previous angular region and the new angular region as shown in Fig. 6.7a. If the Δ expires, then the node sends one or more RREQ with the long hop flag set in each angular region of the list. In a region more than one RREQ are sent if it contains more than one beams. The RREQ packet with the long hop flag set is sent at higher transmit power.

When the destination node receives a RREQ packet targeted to the node, it sends a RREP packet back to the source following the method of the base DSR protocol. An intermediate node at receiving a RREP packet collects all the RREQ from the list of the entry in its PAT whose target address matches with the source address of the RREP packet. It forms a new RREP packet corresponding to each RREQ found in the list and sends the RREP packet back to the source of the RREQ packet. It includes a route in the new RREP packet by concatenating the route found in the RREQ packet from the source of the RREQ to the node itself with the route from itself to the destination. After sending all RREP packets, it purges the entry from the PAT.

Route Maintenance

The route discovery method described above attempts to bridge network partitions that are more permanent in nature. These network partitions would otherwise result in route discovery failure. Similarly, a procedure can be developed for route maintenance to repair a broken route due to node mobility. The main idea of this scheme is that an intermediate performs local route repair after discovering that the next hop node for a packet is unreachable. In general, local repair in a directional antenna system resorts to local search of the unreachable next hop node in other beams. In this case, local repair estimates the direction of the next-to-next hop node and sends the packet through directional transmission to the next-to-next hop node at twice the transmit power normally used for omni-directional transmission. The direction of a next-to-next hop node is estimated considering a parallelogram of the node, next hop node and next-to-next hop node as shown in Fig. 6.7b. The estimate of the direction of vector AC requires the knowledge of the direction of the vectors AB and BC,

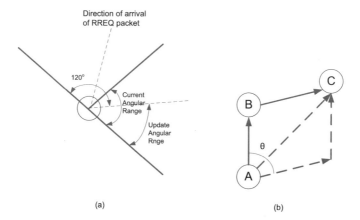

(a)   (b)

Fig. 6.7 (a) Example of updating an angular region. (b) Estimation of the direction to the next-to-next hop node [8]

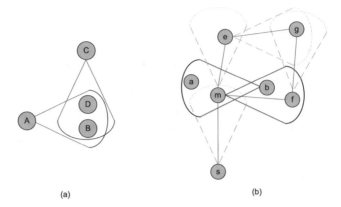

(a)   (b)

Fig. 6.8 (a) Capture prevents parallel communication between A–B and C–D; (b) Example of CaRP where link m-f experiences capture cost [2]

which can be achieved at A if B sends the direction of the vector BC. Node B can send A its direction to C only after gleaning the direction from receiving at least one packet from C. Node A detects that node B is unreachable by counting the failure of earlier transmissions in skip counter it keeps associated with the next hop node B. It then forwards the data packet over the long hop to C through directional transmission at twice the equivalent omni-directional power and setting the long hop flag in the source route header.

6.4.5 Capture-Aware Routing

A node forms specific beam pattern during transmission-reception of packets with its neighbor. During idle state the node listens in all beams for a packet. If it senses an ongoing transmission in a beam, it receives the packet in the MAC layer and decides to either keep or discard the packet based on the destination address, which is known only in the MAC layer. During this period, the node remains captured in a particular beam, as a result remains deaf to transmissions in other beams. Figure 6.8a shows nodes A and C try to transmit simultaneously a packet to nodes B and D respectively. If D forms its beam to As transmission, then it will be captured in that beam and remain deaf to Cs transmission. In [2], Capture-aware Routing Protocol (CaRP) is proposed that defines a routing metric to quantify cumulative effect of Capture in the links of a route. The protocol relies on capture MAC protocol, also proposed in [2].

The Capture MAC protocol defines recurring communication interval of duration T_{cycle}. Each interval is divided into an ON-period and an OFF-period. During the ON-period communication pattern is detected that determines the beamforming pattern β_{off}, which is used to form beams during the subsequent OFF-period. Thus, instead of sensing transmission in every beam during the OFF-period, only beams in the beam pattern β_{off} are activated. All nodes enter ON-OFF periods in a synchronous fashion. During the ON-period, a node keeps track of the packets it receives with the beam-id. It then categorizes the packets into either productive packets or captured packets. Productive packets are the packets that are destined to the node itself, whereas captured packets are the packets that are destined to other nodes. If it receives all captured packets in a beam, then that beam is dropped from the beam pattern β_{off} of that node. If it goes to idle state during the ON-period, it continues listening the channel in omni-directional mode. During the ensuing OFF-period, the node only activates the beams in β_{off} for both sending-receiving and listening in the idle state.

The route discovery may miss a number of beams if it is performed in the OFF-period, since RREQ packets is sent only in the beams given in β_{off}, which results in converging to sub-optimal routes. The CaRP addresses the inefficiency of route discovery in OFF-period by performing route discovery again during the subsequent ON-period, which may lead to finding optimal route. It quantifies the capture cost of a beam B at a node i as the number of neighbors m that capture node i in that beam. Let B_{ij} is the beam at node i to its neighbor j and CB_{ij} be the capture cost of that beam, then the capture cost of the link $i - j$ is given as: $K_{ij} = CB_{ij} + CB_{ji}$. Then, the capture cost of a route R is given as:

$$K_R = \sum_R K_{ij} \tag{6.1}$$

The route cost K_R does not increase monotonically with increasing hop count, since the capture cost of a link may be zero. This can lead to undetected routing loops. Hence, capture cost K_{ij} cannot be used solely as a route metric. It is augmented with cost H_{ij}, such that $\sum H_{ij}$ provides a measure of the hop count of the route R. Further, two routes with the same capture cost and hop count may have different throughput, because the transmission capacities of intermediate nodes in different routes available to the flow are different. The protocol defines participation cost P_i of a node i as a measure of its participation in different flows. It assumes positive value for active nodes that participate in traffic flows, while remains zero for inactive nodes that do not participate in any traffic flow. It is suggested that $P_i = 1$ for source and destination nodes of a flow, and $P_i = 2$ for an intermediate node that carries at least one flow. The unified route metric, U_R, is defined as weighted sum of the three costs:

$$U_R = \sum_{(i,j)\in R} = \omega_k K_{ij} + \omega_P P_i + H_{ij} \tag{6.2}$$

In the above formula, ω_k and ω_p are the weights that control the influence of capture cost and participation cost in the route metric. In Fig. 6.8b, assume node a sends traffic to node b and a route is determined for a new flow from s to g. There are two routes from s to g: R1 (s-m-f-g) and R2 (s-m-e-g). The participation costs of $P_s = P_g = 1$ and $P_m = P_e = P_f = 2$, since none of these nodes participate in any other flow. Let us consider link m-f: when m forms beam towards f, B_{mf}, it may cause b capture m, and similarly a may capture f in the beam B_{fm}. Thus, $CB_{mf} = CB_{fm} = 1$, and the capture cost of the link m-f, $K_{mf} = 2$. Let us assume $H_{ij} = 1$ for every link, and $\omega_k = \omega_p = 1$. The total cost of all the links are as follows: $L_{me} = L_{eg} = L_{fg} = 0 + 2 + 1 = 3$, $L_{mf} = 2 + 2 + 1 = 5$ and $L_{sm} = 0 + 1 + 1 = 2$.

The total cost of routes R1 (s-m-f-g) and R2 (s-m-e-g) are as follows: $U_{R1} = 2 + 5 + 3 = 10$ and $U_{R2} = 2 + 3 + 3 = 8$. In capture-aware routing route R2 is selected for the flow from node s to node g.

References

1. Choudhury R, Vaidya N (2005) Performance of ad hoc routing using directional antennas. Elsevier J Ad Hoc Netw 3(2):157–173
2. Choudhury R, Vaidya N (2007) MAC-layer capture: a problem in wireless mesh networks using beamforming antennas. IEEE sensor, mesh and ad hoc communications and networks (SECON), San Diego, California, pp 401–410
3. Gossain H, Joshi T, Cordeiro C, Agrawal DP (2006) DRP: an efficient directional routing protocol for mobile ad hoc networks. IEEE Trans Parallel Dist Syst 17(12):1438–1541
4. Joshi T, Cordeiro C, Vogety S, Yin J, Gossain H, Agrawal DP (2004) Broadcasting over switched single beam antenna systems. In: IEEE international conference on networks (ICON), pp 671–675
5. Kolar V, Rogers P, Abu-Ghazaleh NB (2005) Route compaction for directional route discovery in MANETs. IEEE international conference on wireless and mobile computing, networking and communications (WiMob), vol 3. Montreal, Canada, pp 101–108
6. Nasipuri A, Castaneda R, Das SR (2001) Performance of multipath routing for on-demand protocols in mobile ad hoc networks. Mobile Netw Appl 6:339–349
7. Roy S, Saha D, Bandyopadhyay S, Ueda T, Tanaka S (2003) A network-aware MAC and routing protocol for effective load balancing in ad hoc wireless networks with directional antenna. In: ACM international symposium on mobile ad hoc networking and computing (MobiHoc), Annapolis, Maryland, pp 88–97
8. Saha AK, Johnson DB (2004) Routing improvement using directional antennas in mobile ad hoc networks. In: IEEE global telecommunications conference (GLOBECOM), vol 5, Dallas, Texas, pp 2902–2908

Chapter 7
Directional QoS Routing Framework

Abstract To cope with the pressing need of running content-rich multimedia applications and real-time services, Quality of Service (QoS) support has become a vital component in today's wireless networks. Using beamforming antennas in multi-hop wireless networks can potentially spare more network resources which can be utilized to provide additional QoS guarantees. However, these opportunities are still under-explored. In this chapter, we develop a framework to analyze conflicts between wireless links in the presence of contention-based directional MAC protocols. Based on a novel taxonomy, we classify the link conflicts and propose a novel colored conflict graph abstraction to model the interdependencies between wireless links. Our analytical framework can be utilized as a basis to design several interference-aware and QoS-aware schemes for contention-based multi-hop wireless networks with beamforming antennas. Based on our analysis, we formulate the bandwidth-guaranteed routing problem as an optimization problem. Since the problem is NP-hard, we present a heuristic algorithm for joint routing and admission control to find single-path bandwidth-guaranteed routes. Using extensive simulations, we demonstrate the accuracy of our conflict analysis and the ability of the proposed algorithm to provide QoS guarantees along with efficient channel utilization.

7.1 QoS Guarantees in Multi-hop Wireless Networks

Quality of Service (QoS) support has become an essential component in today's wireless networks that are designed for running multimedia applications and real-time services. Most of these applications demand QoS guarantees in terms of throughput, delay, jitter or packet loss. Among these requirements, a minimum throughput is the most common since a guaranteed throughput is often needed along with other constraints. In the rest of this chapter, we focus on the bandwidth demand of a flow as its QoS requirement.

Satisfying QoS requirements is often performed by means of admission control and resource reservation. A flow is admitted if the available network resources can provide the required QoS without affecting the QoS of the already admitted flows.

© The Author(s), under exclusive license to Springer Nature Switzerland AG 2021
O. Bazan et al., *Beamforming Antennas in Wireless Networks*,
SpringerBriefs in Electrical and Computer Engineering,
https://doi.org/10.1007/978-3-030-77459-2_7

Upon admission, the network resources should be reserved for the admitted flow. QoS routing is considered a vital part of QoS provisioning in multi-hop wireless networks as it is responsible of determining the path between the source and the destination that can fulfill the QoS requirements. As part of QoS routing, it is necessary to accurately estimate the available network resources along the possible paths and use this information to optimize the path selection process.

Providing QoS guarantees is very challenging in multi-hop wireless networks compared to wired networks [9]. These challenges include the unreliable wireless channel, lack of centralized control, node mobility and channel contention. Due to the shared nature of the wireless medium, it is not an easy task to predict either the available bandwidth before admitting a new flow or the impact of the new flow on the existing flows. Interference between wireless links makes it difficult to get accurate information about the resource availability since it does not always depend only on the local information available at the sender and receiver nodes [22]. The available bandwidth estimation process is highly dependent on the underlying MAC protocol. Hence, QoS routing solutions can be classified as follows [9].

7.1.1 QoS Routing with Contention-Free MAC

In order to avoid possible collisions, contention-free MAC protocols, such as Time Division Multiple Access (TDMA), are sometimes considered since they provide strict scheduling. In TDMA, time is divided into frames which are divided into slots that are used for transmission. Conflict-free schedules are then established for medium access. Although providing optimum schedules is known to be NP-complete, heuristic approaches can achieve sub-optimal performance [16]. QoS routing protocols relying on contention-free MAC protocols can provide pseudo-hard QoS guarantees [9] since they can quantify the resource availability and reservation more accurately. Hence, several QoS routing protocols for ad hoc networks have been proposed in the TDMA environment [2, 13] despite the difficulty to achieve the slot allocation schedules due to the lack of a central controller, node mobility and, more importantly, the complexity and overhead involved [9].

Since the research on QoS routing with beamforming antennas is still in its infancy, the focus was on QoS routing over contention-free MAC protocols as well. In [7], the authors studied the impact of smart antennas on QoS routing in multi-hop wireless networks. The evaluation was done with an extension of a TDMA-based routing algorithm originally proposed for omni-directional antennas. In [8], the authors present a similar link-bandwidth calculation algorithm but no evaluation is done for the proposed algorithm. In [11], a TDMA-based bandwidth reservation protocol is proposed for QoS routing in mobile ad hoc networks with directional antennas. The simulation results clearly show a significant gain in performance relative to the case of omni-directional antennas with an increase in the number of successfully received packets, as well as a decrease in the QoS path acquisition time. In [3], the authors propose a bandwidth-based multipath routing protocol for QoS support in ad hoc networks using the concept of cross-links paths. However, their MAC is Code Division Multiple Access (CDMA) over TDMA which is difficult to implement in a large-scale multi-hop wireless network.

7.1.2 QoS Routing with Contention-Based MAC

As a result of their simplicity, contention-based MAC protocols are commonly used in multi-hop wireless networks. They are mainly based on the concept of CSMA/CA. Accordingly, researchers have shifted their attention to find solutions for QoS routing along with contention-based MAC protocols such as IEEE 802.11 MAC. In a contention environment, it is more complicated to estimate the available resources since CSMA/CA is a non-deterministic medium access. Moreover, the injection of new traffic increases contention in the network which increases the impact of hidden terminals. Hence, the available resources can only be statistically estimated and only soft QoS guarantees can be provided.

A contention-aware admission control protocol for ad hoc networks is designed in [22]. The authors calculate the residual bandwidth based on not only the available local resources at each node but also the resources at the nodes in the carrier-sensing range. In addition, they consider the effect of the contention between nodes along the route used by the flow [23]. However, they ignore the effect of hidden terminal problems. In [14], the work is extended by considering the impact of parallel transmissions in the carrier sensing range to provide a more accurate estimation of the available bandwidth. In [21], the authors develop an analytical model to calculate the capacity of a given path in multi-hop IEEE 802.11 network without violating the QoS of existing flows. The capacity is represented as the reciprocal of the average service time. They consider the different impacts from the carrier sensed traffic and the hidden terminals traffic. Although their proposed model gives valuable insights on how to estimate the residual bandwidth in IEEE 802.11 multi-hop networks, the considered problem is much easier than the QoS routing problem since the path is already given.

The use of beamforming antennas in multihop wireless networks could significantly spare the network resources that can be utilized for additional QoS guarantees. However, in the presence of contention-based directional MAC, existing approaches for QoS provisioning are limited to priority-based services [15, 18, 20]. To provide QoS guarantees, beamforming antennas add unprecedented challenges to the bandwidth estimation process due to directional-specific phenomena such as deafness, directional hidden terminals and MAC-layer capture as discussed in Chap. 3. As a result, it is of great importance to analyze the interdependencies between the wireless links and clearly evaluate the impact of both the physical interference and the contention-based MAC on the link bandwidth in mutli-hop wireless networks with beamforming antennas. Deriving an analytical model to the interference pattern of such networks is indeed essential in the design of efficient QoS-aware routing schemes and this is the focus of the following section.

7.2   Conflict Analysis Framework

In a wireless network, the transmission on a wireless link causes interference on other links within a proximity defined by acceptable signal-to-interference ratio. For a successful transmission to occur on a certain link, interfering links should remain inactive during the transmission period. Hence, interference is a major factor in creating interdependency among wireless links and limiting the capacity of wireless networks. In order to provide guaranteed bandwidth to network flows, interference dictates the constraints that decide which flows are admitted and how they are routed. In addition to their MAC layer overhead, the carrier sensing mechanism in contention-based MAC protocols cannot prevent all types of interference and packet collisions can still occur due to hidden terminals. In the context of beamforming antennas, carrier sensing also fails to prevent transmission failures due to deafness and directional hidden problems. Hence, it is important to analyze and model the interference in a contention-based multi-hop wireless network with beamforming antennas before formulating a QoS routing problem.

7.2.1   System Model

We model the multi-hop wireless network as a directed graph $G = (V, E)$, where V is the set of nodes and E is the set of directed links. The graph G is called the network or connectivity graph. In our network model, we do not assume the traffic demand matrix is known apriori. Instead, we consider a practical traffic demand model in which the traffic flows are generated at different random times. Each flow is specified with random source, random destination and random bandwidth requirement.

Each node is equipped with a switched beam directional antenna consisting of N non-overlapping beams collectively covering all directions [6]. The antenna can operate in one of the two modes, either omni-directional or directional mode, at a time. In the omni-directional mode, the node can receive signals from any direction. Once the signal is sensed, the antenna detects the direction from which the received power is strongest and switches to the directional mode. During transmission, the node operates in the directional mode. Most existing research assumes this antenna model.

Without the loss of generality, we consider the protocol interference model [10] where $R_{Int} = \Delta R_{Com}$ and $\Delta \geq 1$, where R_{Int} is the interference range and R_{Com} is the communication range. In the presence of beamforming antennas, R_{Com} and R_{Int} are no longer represented by the traditional concentric circles but they are function of the beamforming directions of the transmitting and receiving antennas. The communication range in a certain direction can be calculated as

$$R_{Com} = \left[\frac{P_t G_t(\theta_t) G_r(\theta_r)}{K \, \Omega} \right]^{1/\delta},$$

(7.1)

where G_t and G_r are the transmitter and receiver gains, θ_t and θ_r are the transmitting and receiving angles measured with respect to the boresight, δ is the path loss exponent, K is a constant that is a function of the wavelength and Ω is the receiver sensitivity threshold which depends on the acceptable bit-error-rate and the modulation rate.

In our analysis framework, we consider the class of contention-based directional MAC protocols that is based on the concept of CSMA/CA such as [1, 6, 12, 19]. Channel reservation is performed using RTS/CTS handshake before data transmission. Without the loss of generality, the RTS, CTS, DATA and ACK packets are all transmitted directionally. However, idle nodes listens to the medium in omnidirectional mode. Prior to RTS transmission, directional physical carrier sensing is preformed for DIFS period (similar to IEEE 802.11) with the antenna beam directed towards the receiver. When RTS is received, the intended receiver responds with a directional CTS after a SIFS period. The backoff period is performed in an omnidirectional mode to alleviate the possibility of deafness chains and deadlocks [5].

7.2.2 Classification of Link Conflicts

When a contention-based MAC protocol is in operation, the interdependency between the wireless links is a factor not only of the network topology but also of the amount of traffic the links carry since the injection of a new traffic increases the contention in the network. In this section, we study the impact of the topology on the interdependency among wireless links referred to as the interference pattern. In a typical CSMA/CA MAC, e.g. IEEE 802.11, the source performs carrier sensing before transmission to avoid interfering links. If a carrier is detected, the source withholds its transmission and freezes its contention window until the medium is sensed idle. However, interference cannot be completely avoided due to the presence of the hidden terminals that cannot be sensed by the sender but can cause collision at the destination which results in transmission failures. When beamforming antennas are used, link conflicts are significantly increased due to new causes of transmission failures like deafness and directional hidden terminals.

Based on the above discussion, it is evident that for any wireless link in a contention-based multi-hop network with beamforming antennas, there are several types of interfering links with different causes and different impacts on the given link. Figure 7.1 shows our proposed taxonomy of the conflicts between the wireless links. Any conflicting link that interferes with a given link (i, j) in the network graph can be categorized under one of these two major categories: Source-based Conflict (SC) or Destination-based Conflict (DC). An SC link, with respect to a given link, is a link whose transmission interferes with the source of the given link. Similarly, a DC link, with respect to a given link, is a link whose transmission causes interference with the destination of the given link leading to transmission failure on the given link. Within each category, the link conflict could be either strong conflict or weak conflict. By strong conflict, we mean that the two links can never carry successful transmissions

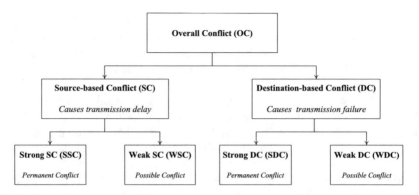

Fig. 7.1 A classification of conflicts between the wireless links in contention-based multi-hop wireless networks with beamforming antennas

concurrently. However, weak conflict refers to the case when the conflicting link may or may not affect the successful transmission over the given link. Weak conflicts are unique features of contention-based multi-hop wireless networks with beamforming antennas. This is mainly because nodes are allowed to transmit directionally but they remain in an omni-directional mode when they are idle. Specific cases of weak conflict links will be discussed later in this section.

Formally, any link (m, n) is a Strong Source-based Conflict (SSC) link with respect to given link (i, j) if:

(a) $m = i$, which means any outgoing link from the source node of the given link.
(b) $n = i$, which means any incoming link to the source node of the given link.
(c) $d(i, m) \leq b_{mn}(i, j).R_{CS}$, where $d(i, m)$ is the distance between node i and node m, $b_{mn}(i, j)$ is a Boolean variable which indicates if the transmission over link (m, n) can be directionally carrier sensed by node i using the same beam used for transmission over link (i, j), and R_{CS} is the carrier sensing range.

The two links cannot carry concurrent transmission if one of the above conditions holds. The transmission on one link should wait for the transmission on the interfering link to complete before starting its own transmission. It is worthy to note that some directional MAC protocols [1, 12] that address the Head-of-Line blocking problem could minimize the effect of the waiting time in case (c) by exploring transmission opportunities over other beams.

Any link (m, n) is a Weak Source-based Conflict (WSC) link with respect to a given link (i, j) if and only if $d(i, m) \leq \overline{b_{mn}}(i, j).R_{CS}$, which means the case of a transmission over link (m, n) that can be carrier sensed with any beam other than the beam used for transmission over link (i, j). Since the node remains in an omni-directional mode during idle time as well as the backoff period, it can be captured by unnecessary transmissions received from different directions delaying the transmission on the given link. In this case, if link (m, n) starts transmission, node i is captured and has to postpone transmission over link (i, j) so the two links cannot be active simultaneously. However, if link (i, j) starts transmission first, link (m, n) can be active at the same time without any impact on link (i, j). Hence, there is a weak conflict relationship between the two links.

A link (m, n) is a Strong Destination-based Conflict (SDC) link of the given link (i, j) if:

(a) $m = j$ and $b_{mn}(i, j) = 0$, which means any outgoing link from the destination node of the given link whose transmission cannot be sensed by the source node.
(b) $n = j$, which means any incoming link to the destination node of the given link.
(c) $d(j, m) \leq b_{mn}(j, i) \cdot R_{Int}$, which means that node j senses the transmission over link (m, n) using the same beam it uses to receive transmission from node i.

Case (a) represents a typical deafness scenario for link (i, j). Case (b) is a deafness scenario if beam $(j, i) \neq$ beam (j, m) but is a hidden link scenario if beam $(j, i) =$ beam (j, m). Case (c) is a hidden link scenario. In all the above cases, the transmission over link (i, j) fails either because the destination is deaf or collision occurs. Hence, for a successful transmission over link (i, j), the two links cannot be active at the same time.

A link (m, n) is a Weak Destination-based Conflict (WDC) link of link (i, j) if and only if $d(j, m) \leq \overline{b_{mn}}(j, i) \cdot R_{Int}$. This is the case where the destination of the given link is captured by another transmission from a direction other than the direction of the source. If transmission over link (m, n) is initiated before the transmission of link (i, j), then node j will appear deaf to node i for the transmission over link (i, j). Conversely, the transmission over link (m, n) has no impact on an already ongoing transmission on link (i, j).

7.2.3 Colored Conflict Graph

Based on the concept of conflict graph [10], we model the interference pattern of the network. A vertex in the conflict graph represents a link in the connectivity graph while an edge in the conflict graph indicates that the two links interfere with each other. Although the impact of link conflicts on the overall capacity is the same when perfect scheduling is employed, this is not the case with contention-based MAC protocols. In order to model these different types of interference, we propose a Colored Conflict Graph (CCG) which is similar to the regular conflict graph except that the directional edges of the new graph have different colors according to the nature of conflict (interference) between the corresponding vertices. Based on the previous categorization, we can build the Colored Conflict Graph (CCG) in which each directed CCG edge has a color that is related to the type of conflict (interference) the end vertex causes on the originating vertex. Figure 7.2 shows an illustrative example. The network graph represents a simple network with 8 nodes each is equipped with 4-beams switched beamforming antenna. For the sake of clarity, we take into consideration the links marked with the solid arrows only. Figure 7.2b shows a portion of

Fig. 7.2 A simple example
to illustrate the colored
conflict graph

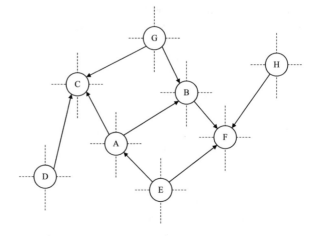

(a) Network graph

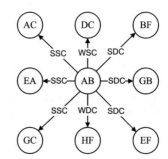

(b) Portion of the corresponding
CCG

the corresponding CCG that illustrates the conflicts with the given link AB. Conflicts
with other links are omitted for clarity. As we can see, link AB has conflicts with
8 other links. Source-based conflicts are those conflicts that affect the source of the
given link AB which is node A. In this example, link AC is SSC link (case (a)),
link EA is SSC link (case (b)) and link GC is SSC link (case (c)). The given link
AB cannot be active simultaneously with any of these links as a result of carrier
sensing. Link DC is considered a WSC link with link AB because DC may capture
node A during its omni-directional listening period preventing it from transmission.
However, links AB and DC can sometimes be active concurrently if the transmission on link AB precedes the transmission on link DC as discussed before. On the
other hand, destination-based conflicts are those conflicts that cannot be sensed by
the source A but prohibits the successful reception at the destination B. Link BF is
SDC link (case (a)), link GB is SDC link (case (b)) and link EF is SDC link (case
(c)). Node B cannot receive A's transmission successfully if any of the above links

are active. On the other hand, link HF is considered a WDC link since the conflict occurs only if the transmission on link HF precedes the transmission on link AB.

We group the interfering links into separate sets, one for each conflict type. For a given link (i, j), we denote the sets of conflict links as $SC(i, j)$, $DC(i, j)$, $SSC(i, j)$, $WSC(i, j)$, $SDC(i, j)$ and $WDC(i, j)$ according to the categories of the conflicts discussed above, where $SC(i, j) = SSC(i, j) \cup WSC(i, j)$ and $DC(i, j) = SDC(i, j) \cup WDC(i, j)$. In addition, we define the set of all conflict links as an overall conflict set, where $OC(i, j) = SC(i, j) \cup DC(i, j)$. Similarly, the strong overall conflict set is given by $SOC(i, j) = SSC(i, j) \cup SDC(i, j)$. Note that, we consider the link (i, j) itself a member of all its conflict sets except the weak conflict sets.

7.3 QoS Routing Problem

Based on the above analysis, it is clear that the interference pattern in multi-hop wireless networks with beamforming antennas is more complicated than their omni-directional counterparts. Using the proposed framework, conflict-aware network mechanisms can be designed for wireless mesh networks with beamforming antennas. For instance, a proper queueing/scheduling mechanism can address Source-based Conflicts (SCs) since their main drawback is the transmission delay. One good example is the Opportunistic Directional MAC protocol (OPDMAC) proposed in [1] which minimizes the idle waiting time by eliminating the need of: (i) An idle defer when the medium is sensed busy in one direction. (ii) An idle backoff period following transmission failures.

In this section, we apply the proposed conflict analysis framework in studying the QoS-aware routing for wireless mesh networks with beamforming antennas. We consider the bandwidth demand of a flow as its QoS requirement. In a contention-based mesh network, a QoS-aware routing should consider both Source-based Conflicts (SCs) and Destination-based Conflicts (DCs). As a result of carrier sensing, a source node can sense SCs but cannot anticipate DCs giving rise to both deafness and directional hidden terminal problems. As discussed before, destination-based conflicts are the sources of transmission failures that triggers packet retransmission which degrades the medium utilization and increases the per-hop delay. Hence, we are interested in maximizing the probability of successful transmission over the wireless link. This metric directly reduces the consequences of DCs because they are the sources of transmission failures. At the same time, we need to reduce the SCs in order to increase the transmission probability of the transmitting node and consequently the probability of successful transmission. Therefore, a conflict-aware routing can exploit the spatial reusability offered by beamforming antennas in wireless mesh networks and, hence, can provide additional QoS guarantees.

7.3.1 Problem Formulation

In this section, we formulate our QoS routing problem. We model the multi-hop wireless network with beamforming antennas as a directed graph $G = (V, E)$, where V is the set of nodes and E is the set of directed links. Given a network topology and the beamforming antenna pattern at each node with information about the existing traffic conditions in the network, we seek a single path (if one exists) between a source-destination pair (s, t) that can support a specified bandwidth requirement when a contention-based directional MAC protocol is in operation. We formulate the problem as follows:

$$\min \sum_{(i,j)\in E} X(i, j) \cdot |\log P_S(i, j)| \tag{7.2}$$

subject to

$$\sum_{(i,j)\in E} X(i, j) - \sum_{(i,j)\in E} X(j, i) = \begin{cases} 1 & i = s, \\ 0 & i \in V \sim \{s, t\} \\ -1 & i = t \end{cases} \tag{7.3}$$

$$0 < P_S(i, j) \le 1 \quad \forall (i, j) \in E, \tag{7.4}$$

$$X(i, j) = 0 \text{ or } 1 \quad \forall (i, j) \in E, \tag{7.5}$$

$$\sum_{(i,j)\in E} X(i, j) \le 1 \quad \forall i \in V, \tag{7.6}$$

where $P_S(i, j)$ is the probability of a successful transmission over link (i, j) given that all the bandwidth requirements are satisfied and $X(i, j)$ is a Boolean variable which has a value of 1 if the route between s and t goes through link (i, j) and 0 otherwise.

In the above formulation, the objective in (7.2) is to minimize the sum of the absolute values of the logarithms of the probabilities of successful transmissions over each link along the chosen path. This is equivalent to maximizing the logarithm of the product of the probabilities of successful transmissions. The constraint in (7.3) is a flow conservation constraint that ensures that the flow will originate from the source s and ends at the destination t. The constraint in (7.4) enforces the probability of successful transmission over each link to be between 0 and 1. This is the bandwidth conservation constraint which ensures that the link capacities are not violated. This constraint also makes sure that the injection of a new flow into the network will not violate the bandwidth-guarantee for existing flows. The constraints in (7.5) and (7.6) enforce the single path restriction on the flow from s to t.

Theorem 7.1 *The QoS routing optimization problem is NP-hard.*

Proof The corresponding decision problem of our optimization problem is the bandwidth-constrained routing problem in wireless networks which is proven to be NP-complete in [4]. Hence, it follows by definition of NP-hardness that our QoS routing optimization problem is NP-hard.

Since the above formulation is hinged on the probability of a successful transmission over each wireless link, the key to solve the problem is to derive a closed form expression for $P_S(i, j)$, which is presented in the next subsection.

7.3.2 Probability of Successful Transmission $P_S(i, j)$

In contention-based multi-hop wireless networks, the probability that a transmission is successful depends mainly on the contention which is related to both the interference pattern and the traffic rates. In Sect. 7.2.3, we modeled the interference pattern using CCG. To incorporate the effect of the ongoing traffic, we rely on the link utilization ρ. The link utilization is the fraction of link capacity that is used by the ongoing traffic. If the average traffic load over a link (i, j) is R packets/sec, the link utilization $\rho(i, j)$ can be expressed as:

$$\rho(i, j) = R \cdot T_{service} \tag{7.7}$$

where $T_{service}$ is the average service time needed to transmit a data packet with an average size of L bits. Due to the MAC layer overhead, the average service time $T_{service}$ can be expressed as:

$$T_{service} = T_{RTS} + T_{CTS} + \frac{L + H}{C} + T_{PLCP} + T_{ACK} + T_{DIFS} + 3T_{SIFS} + T_{BO} \tag{7.8}$$

where L is the packet data size, H is the MAC header length, C is the channel capacity in bits/sec, T_{RTS}, T_{CTS} and T_{ACK} are the transmission times of RTS, CTS and ACK packets respectively. T_{PLCP} is the transmission time of the Physical Layer Convergence Procedure (PLCP) header. The terms T_{DIFS} and T_{SIFS} represent the inter-frame spaces and he term T_{BO} denotes the average time spend during the backoff period. Since our goal is to maximize the probability of successful transmission, packet retransmissions are expected to be negligible [14, 22].

Once link utilization of conflicting links in the network are known, the probability of a successful transmission over link (i, j), $P_S(i, j)$, can be expressed as:

$$P_S(i, j) = 1 - P_F(i, j) = 1 - \frac{\rho_{DC(i,j)}}{1 - \rho_{SC(i,j)}} \tag{7.9}$$

where $P_F(i, j)$ is the probability of a failed transmission over link (i, j) and $\rho_{DC(i,j)}$ and $\rho_{SC(i,j)}$ represent the aggregate link utilization of all the links in the sets $DC(i, j)$ and $SC(i, j)$ respectively. Since the term $1 - \rho_{SC(i,j)}$ represents the link availability at the source i, the term $P_F(i, j)$ is equivalent to the ratio between the fraction of time in which the destination j cannot receive a successful transmission from i and the fraction of time the source i is allowed to transmit.

In (7.9), $P_S(i, j)$ does not consider that the transmissions over SC-links and DC-links can overlap in time due to the possible spatial reuse achieved by beamforming antennas. Hence, an admission control that is based on (7.9) would be over-conservative. To take the spatial reuse into account, the probability of a successful transmission over link (i, j), $P_S^{SR}(i, j)$, can be expressed as:

$$P_S^{SR}(i, j) = 1 - \frac{\rho_{OC(i,j)} - \rho_{SC(i,j)}}{1 - \rho_{SC(i,j)}} \tag{7.10}$$

where $\rho_{OC(i,j)}$ represent the aggregate link utilization of all the links in the set $OC(i, j)$.

The aggregate link utilization of all the links in a conflict set XY (where XY is $SC(i, j)$ or $DC(i, j)$) or $OC(i, j)$)can be calculated as:

$$\rho_{XY(i,j)} = \rho\left(\bigcup_{(u,v) \in XY} (u, v) \right) \tag{7.11}$$

Due to the fact that some of the links in any conflict set XY may be active concurrently (if they are not in strong conflict with each other), the overall utilization of the union operation in (7.11) is not simply equivalent to the summation of the individual link utilization but the overlapping utilization should be deducted. Since the transmissions of non-interfering links are generally independent, the intersection of their utilizations is equivalent to the product of the individual link utilization. The intersection between the utilization of any group G_r of links is

$$\rho\left(\bigcap_{(u,v) \in G_r} (u, v) \right) = \begin{cases} \displaystyle\prod_{(u,v) \in G_r} \rho(u, v) & (u_1, v_1) \notin SOC(u_2, v_2) \\[2ex] & \forall (u_1, v_1), (u_2, v_2) \in G_r \\[1ex] 0 & \text{otherwise} \end{cases} \tag{7.12}$$

7.3.3 QoS Routing Algorithm

Based on the proposed conflict analysis, we formulated the QoS routing problem as a mixed integer nonlinear programming optimization problem and proved that it is NP-hard. In this subsection, we propose a heuristic to solve it. Our algorithm computes a bandwidth-guaranteed route that maximizes the probability of successful transmission over all the links in a path. The algorithm, which is based on modifications of Dijkstra's algorithm, provides QoS support to the new flow without violating the QoS of the existing flows. A flow should not be admitted if a feasible route is not found.

The goal of our proposed algorithm is to minimize the sum of the weights of the links along the path between a source s and a destination t with a bandwidth requirement B, when a link weight $W(i, j)$ is associated with each link (i, j). We define the link weight $W(i, j)$ as follows:

$$W(i, j) = \begin{cases} |\log P_S(i, j)| & 0 < P_S(i, j) \leq 1 \\ \infty & \text{otherwise} \end{cases} \quad (7.13)$$

where $P_S(i, j)$ is computed using (7.9) or (7.10).

The calculation of the probability of successful transmission over a link takes into consideration the intra-flow conflicts. This kind of conflict cannot be measured before the flow is admitted. Therefore, its impact should be estimated before the admission control decision is taken. For instance, $P_S(i, j)$ at any intermediate hop is calculated assuming there is a virtual flow of bandwidth B on the shortest path from the source s to the intermediate node i. This accounts for the impact of the same flow going on the previous hops. Moreover, if node j is not the destination, we consider the impact of additional bandwidth B on one of j's outgoing links. Based on this definition, it is obvious that our algorithm, in the absence of other traffic, is going to choose the minimum-hop path since each additional intermediate hop will have a weight greater than zero due to the intra-flow deafness. Moreover, this guarantees loop-free routing.

To avoid violation of QoS for the existing traffic flows, we have to recheck the weights of all other links that link (i, j) interferes with assuming that bandwidth B is allocated on link (i, j). If any of these weights have become equal to infinity, this means injecting a traffic rate B on link (i, j) will cause a QoS violation of an existing traffic. Hence, link (i, j) should not be used directly and $W(i, j)$ is set to infinity. Our proposed algorithm is presented in Algorithm 1.

7.4 Performance Evaluation

In this section, we evaluate the accuracy of our analysis and the performance of our proposed QoS routing algorithm via simulations. The simulations are conducted using OPNET simulator. We consider a static wireless network with 20 nodes randomly positioned in an area of $1000\,m \times 1000\,m$. Each node is equipped with a switched beam directional antenna of six non-overlapping beams. The transmission range is set to 250 m. The OPDMAC protocol [1] is the MAC layer protocol, the channel data rate is 11 Mbps and the packet size is set to 1024 bytes. We implemented, in the OPNET simulator, the proposed QoS routing algorithm with admission control

---

Algorithm 1 Routing and admission control algorithm

---

for each node $v \in V$ **do** {/*Initializations*/}
  $W(v) = \infty$ {/*Sum of link weights from the source*/}
  $Path\_to(v) = \{\}$ {/*List of path links from the source*/}
end for
$W(s) = 0$
$Q = V$ {/*The set of unvisited nodes*/}
while $t \in Q$ **do** {/*Main loop*/}
  choose node $i \in Q$ such that $W(i)$ is minimum
  $Q = Q \sim \{i\}$
  for each node $j \in$ neighbors (i) **do**
    calculate $W(i, j)$ according to Eq. (7.13)
    if $W(i, j) < \infty$ **then** {/*link can support QoS*/}
      for each link (m, n) such that $(i, j) \in OC(m, n)$ **do**
        calculate $W(m, n)$ {/*check QoS of other links*/}
        if $W(m, n) = \infty$ **then** {/*QoS violation*/}
          $W(i, j) = \infty$
        end if
      end for
    end if
    if $W(i, j) < \infty$ **then** {/*flow can be admitted on this link*/}
      if $W(j) > W(i) + W(i, j)$ **then**
        $W(j) = W(i) + W(i, j)$ {/*update the total weight of j*/}
        $Path\_to(j) = Path\_to(i) + \{(i, j)\}$ {/*update the path to j*/}
      end if
    end if
  end for
end while
if $W(t) < \infty$ **then**
  Admit the flow and route it along $Path\_to(t)$
else
  Block the flow
end if

---

as well as the maximally zone-disjoint routing algorithm [17] for the sake of comparison since there is no directional routing protocol in the literature that provides bandwidth guarantees. Moreover, the zone-disjoint routing algorithm aims to avoid route coupling.

First, we consider a simple scenario to illustrate the effectiveness of the admission control provided by our algorithm. Six arbitrary Constant Bit Rate (CBR) flows are ready for admission in the network. Each flow request is assigned a random source and a random destination. The starting time and the bandwidth requirement of each flow are shown in Table 7.1.

In Figs. 7.3 and 7.4, we show the throughput and delay of the 6 flows when the zone-disjoint routing algorithm is used. This is the case of no admission control. As expected, the network becomes congested after flow 3 has started. As more flows are injected, the throughput of all flows shows a significant decrease and none of the flows could meet its bandwidth requirement. Once congestion occurs, the

Table 7.1 Traffic configurations

Flow number	1	2	3	4	5	6
Starting time (s)	5	10	15	20	25	30
Bandwidth requirement (Kbps)	195.1	413.9	1316.1	667.4	1259.1	579.9

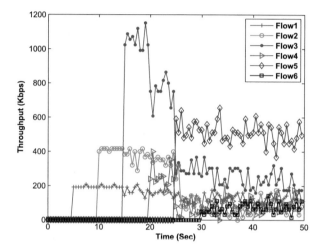

Fig. 7.3 Throughput in case of zone-disjoint routing algorithm (no admission control)

delay experienced by the flows increases dramatically with large variations. Such performance is unacceptable for real-time and multimedia applications.

Figures 7.5 and 7.6 show the throughput and delay of the admitted flows when the proposed QoS routing algorithm is used. Our algorithm admits four flows only to avoid congestion. The admitted flows experience consistent throughput which matches the bandwidth requirement. The worst delay of the admitted flows is about 13 ms which is extremely small and about 1000 times smaller than the worst delay in the case of no admission control shown in Fig. 7.4. This experiment demonstrates that the benefits of admission control when integrated with routing in supporting QoS requirements.

In the next experiment, we verify the accuracy of our conflict analysis and the estimation of the link bandwidth utilization. The QoS routing algorithm should guarantee the QoS requirement of the admitted flows and at the same time should not be over-conservative in rejecting flows that the network can support; otherwise, such rejection may lead to bandwidth under-utilization. We compare the performance of the proposed QoS algorithm under two different QoS metrics. When the spatial reuse is not considered, $P_S(i, j)$ is computed using (7.9) and we refer to this variant of

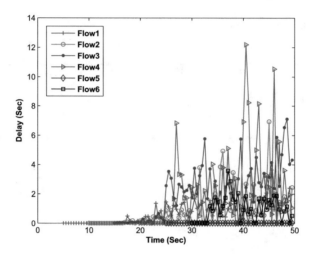

Fig. 7.4 End-to-end delay in case of zone-disjoint routing algorithm (no admission control)

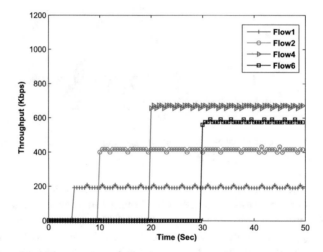

Fig. 7.5 Throughput in case of the proposed QoS routing algorithm (with admission control)

the algorithm as QoS (no-SR). The other variant is refered to as QoS (SR-aware) in which $P_S(i, j)$ is computed using (7.10) to take the spatial reuse feature of beamforming antennas into consideration. We also compare the QoS algorithms with both maximally zone-disjoint algorithm and Dijkstra algorithm with minimum hop-count metric. We consider random networks each with 20 nodes. For each simulation run, 10 CBR connection requests are generated, each with random source, random destination and random bandwidth requirement uniformly distributed between $[0, B_{max}]$.

Figure 7.7 depicts the Packet Delivery Ratio (PDR) under different traffic loads. The PDR is defined as the ratio between the total throughput to the total traffic

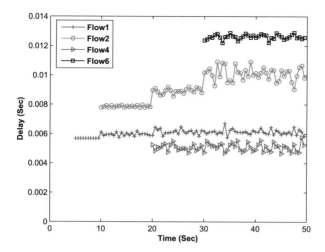

Fig. 7.6 End-to-end delay in case of the proposed QoS routing algorithm (with admission control)

generation rate of the CBR sources. As we can see, both QoS routing metrics achieves a PDR very close to 1. This indicates the ability of the routing algorithms to satisfy the QoS guarantees. In the case of zone-disjoint and minimum hop count metrics, the PDR is less than 1 and decreases as the traffic load increases.

Figure 7.8 shows the total network throughput under different traffic load conditions. As we can see, QoS (SR-aware) provides significant increase in the total throughput when compared QoS (no-SR). When the spatial reuse is considered in the computation of the link utilizations, there is up to 28% increase in the total throughput that can be allowed without compromising the QoS guarantees. This is mainly because QoS (SR-aware) attempts to make full use of the possible spatial reuse by admitting more flows as shown in Table 7.2. From Fig. 7.8, it is also evident that QoS (SR-aware) helps attaining higher throughput when compared with routing algorithms that do not employ admission control. The only exception is when the maximum bandwidth requirement B_{max} is equal to 1400 Kbps. In this case, there is a slight reduction in the total throughput with respect to the minimum hop count metric which is due to the coarseness of the bandwidth requirements. In this experiment, our results show that QoS (SR-aware) is able to effectively utilize the capacity of the network while QoS (no-SR) is over-conservative as expected.

In Fig. 7.9, we present the average end-to-end delay of the admitted flows. The flows experience very small delay under the QoS algorithms as compared to other algorithms indicating the ability of the proposed algorithms to avoid network congestions. In addition to providing bandwidth guarantees, our QoS routing metric is able to provide very small end-to-end delay, in terms of milliseconds, which is desirable for real-time applications.

Note that the success of our QoS algorithm is not only because of the admission control but also due to a better routing metric. As we can see in Table 7.2, all the flows

Fig. 7.7 Average packet
delivery ratio for different
B_{max}

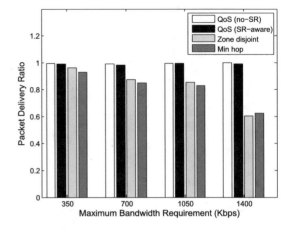

Fig. 7.8 Total throughput
for different B_{max}

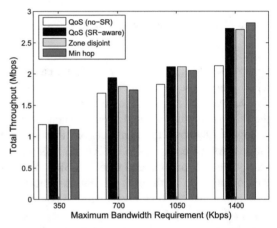

Fig. 7.9 Average delay for
different B_{max}

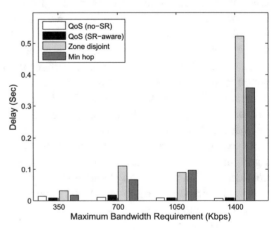

Table 7.2 Percentage of admitted flows by our admission control scheme

B_{max} (Kbps)	350	700	1050	1400
no-SR (%)	100	82	82	54
SR-aware (%)	100	96	90	64

are admitted in the case of $B_{max} = 350$ Kbps, which is the same as other algorithms with no admission control. Hence, our results in this case illustrate the effectiveness of the proposed routing metric as the admission control did not exclude any flow from being scheduled. It is obvious from the figures that the proposed QoS routing algorithm outperforms the other algorithms in terms of the packet delivery ratio, the total throughput and the average end-to-end delay when the same flows are admitted.

References

1. Bazan O, Jaseemuddin M (2011) On the design of opportunistic MAC protocols for multi-hop wireless networks with beamforming antennas. IEEE Trans Mobile Comput 10(3):305–319
2. Chen S, Nahrstedt K (1999) Distributed quality-of-service in ad-hoc networks. IEEE J Sel Areas Commun 17(8):1488–1505
3. Chen Y, Jan S, Chuang M (2007) A shoelace-based QoS routing protocol for mobile ad hoc networks using directional antenna. In: IEEE region 10 conference (TENCON), Taipei, Taiwan, pp 1–4
4. Chiu C, Kuo Y, Wu EH, Chen G (2008) Bandwidth-constrained routing problem in wireless ad hoc networks. IEEE Trans Parallel Distrib Syst 19(1):4–14
5. Choudhury R, Vaidya N (2004) Deafness: a MAC problem in ad hoc networks when using directional antennas. In: IEEE international conference on network protocols (ICNP), Berlin, Germany, pp 283–292
6. Choudhury R, Yang X, Ramanathan R, Vaidya N (2002) Using directional antennas for medium access control in ad hoc networks. In: ACM international conference on mobile computing and networking (Mobicom), Atalanta, Georgia, pp 59–70
7. Coletti L, Ciglioni D, Capone A, Zambardi M (2005) Impact of smart antennas on QoS routing for multi-hop wireless networks. In: IST mobile and wireless communications summit, Dresden, Germany
8. Hamdaoui B, Ramanathan P (2005) Link-bandwidth calculation for QoS routing in wireless ad-hoc networks using directional communications. IEEE international conference on wireless networks, communications and mobile computing (IWCMC), vol 1. Maui, Hawaii, pp 91–94
9. Hanzo L, Tafazolli R (2007) A survey of QoS routing solutions for mobile ad hoc networks. IEEE Commun Surv Tutor 9(2):50–70
10. Jain K, Padhye J, Padmanabhan V, Qiu L (2003) Impact of interference on multihop wireless network performance. In: ACM international conference on mobile computing and networking (MOBICOM), San Diego, California, pp 66–80
11. Jawhar I, Wu J (2006) Resource allocation in wireless networks using directional antennas. In: IEEE international conference on pervasive computing and communications (Percom), Pisa, Italy, pp 318–327
12. Kolar V, Tilak S, Abu-Ghazaleh NB (2004) Avoiding head of line blocking in directional antenna. In: IEEE international conference on local computer networks (LCN), Zurich, Switzerland, pp 385–392

13. Lin C, Liu J (1999) QoS routing in ad hoc wireless networks. IEEE J Sel Areas Commun 17(8):1426–1438
14. Luo L, Gruteser M, Liu H, Raychaudhuri D, Huang K, Chen S (2006) A QoS routing and admission control scheme for 802.11 ad hoc networks. In: ACM workshop on dependability issues in wireless ad hoc networks and sensor networks, New York, USA, pp 19–28
15. Man H, Li Y, Zhuang X (2006) Video transport over multi hop directional wireless networks. In: IEEE international conference on multimedia and expo (ICME), Toronto, Ontario, pp 1525–1528
16. Ramanathan R (1999) A unified framework and algorithm for channel assignment in wireless networks. Wirel Netw 5(2):81–94
17. Roy S, Saha D, Bandyopadhyay S, Ueda T, Tanaka S (2003) A network-aware MAC and routing protocol for effective load balancing in ad hoc wireless networks with directional antenna. In: ACM international symposium on mobile ad hoc networking and computing (MobiHoc), Annapolis, Maryland, pp 88–97
18. Saha D, Roy S, Bandyopadhyay S, Ueda T, Tanaka S (2004) A distributed feedback control mechanism for priority-based flow-rate control to support QoS provisioning in ad hoc wireless networks with directional antenna. In: IEEE international conference on communications (ICC), vol 7. Paris, France, pp 4172–4176
19. Takai M, Martin J, Ren A, Bagrodia R (2002) Directional virtual carrier sensing for directional antennas in mobile ad hoc networks. In: ACM international symposium on mobile ad hoc networking and computing (MobiHoc), Lausanne, Switzerland, pp 183–193
20. Ueda T, Tanaka S, Saha D, Roy S, Bandyopadhyay S (2004) A priority-based QoS routing protocol with zone reservation and adaptive call blocking for mobile ad hoc networks with directional antenna. In: IEEE global telecommunications conference (GLOBECOM) workshops, Dallas, Texas, pp 50–55
21. Wang K, Yang F, Zhang Q, Xu Y (2007) Modeling path capacity in multi-hop IEEE 802.11 networks for QoS services. IEEE Trans Wirel Commun 6(2):738–749
22. Yang Y, Kravets R (2005) Contention-aware admission control for ad hoc networks. IEEE Trans Mobile Comput 4(4):363–377
23. Yin S, Xiongy Y, Zhang Q, Lin X (2006) Traffic-aware routing for real-time communications in wireless multi-hop networks. Wirel Commun Mobile Comput 6(6):825–843

Chapter 8
Open Research Areas

Abstract Directional antenna can boost the performance of wireless networks in terms of coverage, connectivity, spatial reuse and capacity compare to the omni-direction antenna. To meet the ever- increasing demands for high data rate; industry, academia and research groups are investigating several areas of wireless technology to improve the capacity of the system. Directional beamforming antennas, directional multi-hop routing, quality of service aware support and the use of millimeter wave spectrum are some of the key areas received considerable attention in the research community. Although all of the above-mentioned technologies provide potential improvement, they also have technical challenges to address for further improvement. In this chapter we present some open issues and research directions for future generation WLAN systems.

8.1 Cross Layer Design for Directional MAC and Routing

The research work done so far has focused on dealing with the beamforming-related challenges separately at different layers of the protocol stack, namely the MAC and network layers. Nevertheless, many issues such as deafness and hidden terminal problems can impact not on the MAC layer but the routing performance as well. Moreover, certain design choices at one layer can have a severe impact on another layer. For example, forming a routing path in a straight line can introduce additional interference between successive hops degrading the performance at the MAC layer [5]. In order to fully exploit the benefits of beamforming antennas in multi-hop wireless networks, the joint design of directional MAC and routing protocols is needed.

© The Author(s), under exclusive license to Springer Nature Switzerland AG 2021   121
O. Bazan et al., *Beamforming Antennas in Wireless Networks*,
SpringerBriefs in Electrical and Computer Engineering,
https://doi.org/10.1007/978-3-030-77459-2_8

8.2   Multi-hop Wireless Networks with Heterogeneous Antennas

Due to the vast spread of wireless devices with omni-directional antennas, it is economically infeasible to replace all the existing antennas with the new beamforming antenna technology. Moreover, it is sometimes difficult to deploy beamforming antennas except on some special nodes in the network due to practical limitations. Hence, the incremental deployment of beamforming antennas would be the viable way in exploiting their benefits. Recent studies [2, 4, 10] have shown that a wireless networks with a heterogeneous antennas outperform the homogenous omni-directional networks even with a few fraction of directional nodes. Hence, it is crucial to develop heterogeneity-aware MAC and routing protocols for multi-hop wireless networks with heterogeneous antennas. In that direction, interested research may investigate the co-existence between directional MAC and IEEE 802.11 which may lead to heterogeneity-aware MAC-layer enhancements that can be easily implemented to improve the overall performance of heterogeneous networks. Moreover, heterogeneity-aware routing metrics are needed.

8.3   Using Beamforming Antennas in Indoor Environments

The use of wireless communications in an indoor environment usually suffers from multipath fading and scattering which could introduce new challenges to beamforming antennas. In a multipath environment, signals transmitted by neighboring nodes can be received from several directions and may interfere with ongoing directional communications. It is indeed challenging to perform optimum beamforming that can maximize the Signal to interference and noise ratio using simple and time-efficient DSP algorithms. On the other hand, the presence of multiple paths between a transmitter-receiver pair could allow nodes outside their ideal communication region to learn about the ongoing communication and hence transmission failures due to deafness can be reduced. This research area is still underexplored and thus provide numerous opportunities for future research.

8.4   Analytical Modeling for Directional MAC

The majority of the performance evaluations for MAC protocols in multi-hop wireless networks with beamforming antennas were done via discrete event simulations. The main drawback of this evaluation tool is the huge simulation time than limits the scalability of the considered scenarios. Although there are several analytical models for the IEEE 802.11 DCF MAC with the implicit assumption of using omni-directional antennas, very few attempts were made towards the analytical modeling of directional MAC protocols [1, 3, 7, 11]. These attempts have relied heavily on the use of Markov chains with simplistic assumptions regarding the antenna radiation pattern, the physical parameters of the channel and/or the traffic characteristics. Moreover, some antenna-specific MAC challenges, such as deafness, are usually ignored in most of the existing models. Further research need to be conducted to develop more accurate and generic analytical models for MAC protocols in multi-hop wireless networks with beamforming antennas.

8.5 QoS-Aware Directional Routing Protocols

In Chap. 7, we presented a comprehensive analysis of the interdependencies of wireless links in contention-based multi-hop wireless networks with beamforming antennas. Our analysis framework paves the way for developing efficient QoS routing protocols over directional contention-based MAC protocols. We proposed a centralized algorithm that assumes global information is present. However, our problem formulation is amenable to distributed algorithms. Although there are some interesting distributed QoS routing protocols [8, 12] in the literature for the case of omni-directional antennas, the implementation of QoS-aware routing protocols in multi-hop wireless networks with beamforming antennas is indeed challenging and still an open area for research.

8.6 Joint Optimization of Different Physical Layer Capabilities

In this book, we focused on exploiting a single physical layer capability which is the beamforming antennas in order to improve the performance of multi-hop wireless networks. By controlling several physical-layer parameters jointly and adaptively, one can achieve further performance improvement. For instance, by combining beamforming antennas with power control, interference could be significantly reduced resulting in additional spatial reuse of the channel [9]. Other capabilities that can also be jointly considered with beamforming antennas include adaptive modulation, multi-channel transmissions and MIMO techniques.

8.7 Research Issues for mmWave WLAN

Millimeter wave potentially offer many advantages but at the same time opens several research issues for further advancement. Inter AP cooperation, dual connectivity, multiuser MIMO and group beamforming are some key issues for future mmWave

WLAN technologies [13]. If a station suffers poor link or blockage occurs with serving AP, cooperative AP can help the station for better link. Thus, to realize the AP cooperation fast handoff, fast link recovery and joint transmission are needed further research. Dual connectivity is another area of research for next generation WLAN. It can help to support for high frequency and low frequency cooperation. Moreover, if a station stay connected with two APs, when the station moves out of the coverage of one station can get uninterrupted service from another AP. Multiuser MIMO (MU-MIMO) or Massive MIMO still an open issue for further improvement of WLAN. Since directional transmission in mmWave transmission reduce interference MIMO technology could be feasible. Finally, group beamforming can reduce the signalling and feedback overhead. Therefore, creating efficient beam grouping is another highly interested research area [6, 13].

References

1. Alawieh B, Assia C, Mouftah H (2009) Power-aware ad hoc networks with directional antennas: models and analysis. Elsevier J Ad Hoc Netw 7(3):486–499
2. Bazan O, Jaseemuddin M (2010) On the capacity of multi-hop wireless networks with heterogeneous antennas. In: IEEE VTC-Fall
3. Bazan O, Jaseemuddin M (2010) Performance analysis of directional CSMA/CA in the presence of deafness. IET Commun 4(18):2252–2261
4. Beygelzimer A, Kershenbaum A, Lee KW, Pappas V (2008) The benefits of directional antennas in heterogeneous wireless ad-hoc networks. In: IEEE MASS, pp 442–449
5. Choudhury R, Vaidya N (2005) Performance of ad hoc routing using directional antennas. Elsevier J Ad Hoc Netw 3(2):157–173
6. Ghasempour Y, da Silva CR, Cordeiro C, Knightly EW (2017) IEEE 802.11 ay: next-generation 60 GHz communication for 100 Gb/s Wi-Fi. IEEE Commun Mag 55(2):186–192
7. Hsu J, Rubin I (2006) Performance analysis of directional CSMA/CA MAC protocol in mobile ad hoc networks. In: IEEE international conference on communications (ICC), vol 8. Istanbul, Turkey, pp 3657–3662
8. Luo L, Gruteser M, Liu H, Raychaudhuri D, Huang K, Chen S (2006) A QoS routing and admission control scheme for 802.11 ad hoc networks. In: ACM workshop on dependability issues in wireless ad hoc networks and sensor networks, New York, USA, pp 19–28
9. Ramanathan R (2004) Antenna beamforming and power control for ad hoc networks. In: Mobile ad hoc networking. Wiley-IEEE Press, pp 139–174
10. Sundaresan K, Sivakumar R (2006) Ad hoc networks with heterogeneous smart antennas: performance analysis and protocols. Wirel Commun Mobile Comput 6(6):893–916
11. Wang Y, Garcia-Luna-Aceves JJ (2003) Collision avoidance in single-channel ad hoc networks using directional antennas. In: IEEE international conference on distributed computing systems (ICDCS), Providence, Rhode Island, pp 640–649
12. Yang Y, Kravets R (2005) Contention-aware admission control for ad hoc networks. IEEE Trans Mobile Comput 4(4):363–377
13. Zhou P, Kaijun C, Xiao H, Xuming F, Yuguang F, Rong H, Yan L, Yanping L (2018) IEEE 802.11 ay-based mmWave WLANs: design challenges and solutions. In: IEEE Commun Surv Tutor 20(3):1654–1681

Printed in the United States
by Baker & Taylor Publisher Services